AF713737

Environmental Impact of ENERGY STRATEGIES within the EEC

Published for the Commission of the European Communities,
Directorate-General Scientific and Technical Information
and Information Management, Luxembourg

LEGAL NOTICE

Neither the Commission of the European Communities nor any person acting on behalf of the Commission is responsible for the use which might be made of the following information

Environmental Impact of

ENERGY STRATEGIES

within the EEC

A report prepared for the Environment and Consumer Protection Service of the Commission of the European Communities

by

Environmental Resources Limited

Published for the
COMMISSION OF THE EUROPEAN COMMUNITIES
by
PERGAMON PRESS
OXFORD · NEW YORK · TORONTO · SYDNEY · PARIS · FRANKFURT

U.K.	Pergamon Press Ltd., Headington Hill Hall, Oxford OX3 0BW, England
U.S.A.	Pergamon Press Inc., Maxwell House, Fairview Park, Elmsford, New York 10523, U.S.A.
CANADA	Pergamon of Canada, Suite 104, 150 Consumers Road, Willowdale, Ontario M2J 1P9, Canada
AUSTRALIA	Pergamon Press (Aust.) Pty. Ltd., P.O. Box 544, Potts Point, N.S.W. 2011, Australia
FRANCE	Pergamon Press SARL, 24 rue des Ecoles, 75240 Paris, Cedex 05, France
FEDERAL REPUBLIC OF GERMANY	Pergamon Press GmbH, 6242 Kronberg-Taunus, Pferdstrasse 1, Federal Republic of Germany

Copyright © 1980 ECSC, EEC, EAEC, Luxembourg

All Rights Reserved. No part of this publication may be reproduced, stored in a retrieval system or transmitted in any form or by any means: electronic, electrostatic, magnetic tape, mechanical, photocopying, recording or otherwise, without permission in writing from the copyright holders.

First edition 1980

British Library Cataloguing in Publication Data

Environmental Resources Limited
The environmental impact of energy strategies within the EEC.
1. Energy policy - European Economic Community countries 2. Man - Influence on nature - European Economic Community countries
I. Title II. Commission of the European Communities
301.31 HD9502.E862 79-41453

ISBN 0-08-025681-3

EUR 6571 EN

In order to make this volume available as economically and as rapidly as possible the authors' typescripts have been reproduced in their original forms. This method unfortunately has its typographical limitations but it is hoped that they in no way distract the reader.

Printed in Great Britain by A. Wheaton & Co. Ltd., Exeter

CONTENTS

PREFACE

This study was commissioned in 1977 by the European Commission, the principal findings of which were presented in January 1978 to the second session of the EEC Energy Commission's Nuclear Hearings held in Brussels. The final report was delivered in March 1978.

Section 1 of the report considers the requirements of energy of the nine Member States from 1976-2000, applying a detailed examination of sectoral needs under different assumptions of economic growth and levels of energy conservation. The section then goes on to develop alternative energy supply strategies which take into account feasible maximum (and minimum) contributions that could reasonably be expected from different energy sources, bearing in mind the availability of foreign sources of supply and the need in particular to minimise foreign oil imports.

Section 2 initially describes the local, regional and global environmental impacts associated with the production and use of different energy forms. It then goes on to make a quantitative assessment of the impacts associated with each of the alternative energy strategies developed in Section 1. Finally the report puts forward recommendations on Community energy policies designed to minimise overall environmental impact.

Two general points should be made.

First, the study does not consider occupational risks and hazards to plant workers. For example, in nuclear power plants, low level radiation levels are not discussed. This is a separate field of study.

Secondly, although the main body of the research work was largely undertaken in the second half of 1977, the findings and conclusions of the study have not been altered in any significant way by developments since that time. Footnotes have been added where more recent events, information and references are considered to be of significance.

SUMMARY

1. Approach

To evaluate the environmental impact of alternative energy strategies, it was necessary to adopt a three-stage quantitative analysis involving:-

i. forecasts of energy requirements on a detailed sector by sector basis within the Community under different assumptions for future economic growth, the price of energy and the level of conservation;

ii. development of alternative energy supply strategies to meet forecast energy requirements, having in mind their availability and the potential rates of development;

iii. a quantitative assessment of the various environmental impacts deriving from the alternative energy strategies.

Based on iii, recommendations on energy policies were drawn up designed to minimise the overall environmental impact in meeting future energy requirements of Member States.

2. Major conclusions

2.1 Depending on the degree of active energy conservation policies adopted, 1976-2000, primary energy consumption is forecast as growing at 1.8-2.25% per annum. This represents an increase in consumption over the period of 55-79% from 969 million tonnes oil equivalent to 1493-1766 million tonnes oil equivalent. Maximum variations in the 2000 energy consumption forecasts according to different forecast rates of economic growth are estimated as -15% and +10%. In all the strategies and cases examined, the rate of energy consumption is forecast to slow after 1985, partly in response to the increase in energy prices expected for this period.

2.2 Scope for energy conservation varies considerably from around 13-16% in transport, 18-25% in industry sectors, and up to around 40% in the domestic sector. If conservation is left solely to consumer response to higher energy prices, a 12-15% savings in energy consumption may be achieved by 2000; by adopting suitable fiscal policies, legislation,R and D, etc., a further 10-12% in energy savings could be achieved.

2.3 The growth of electricity consumption has been associated with very marked impacts upon the environment. Even in the minimum electricity growth strategy, it is estimated that consumption will rise by a factor of just over 2; in the higher nuclear strategies this increase is estimated as 2.9 times.

2.4 Having in mind likely availability of indigenous resources and the probable limits to imported oil supplies, hydrocarbons are likely to be able to contribute 51-59% of energy supplies by 2000. Even assuming the most optimistic rate of development, renewable energy sources are unlikely to be able to supply more than around 9% of total primary energy requirements by that time, including hydro-electricity contribution of 2%. Maximum indigenous coal production is likely to provide some 12-13% of total energy needs. The critical long term issue on energy policy for the Community would, therefore, appear to be the choice between the level of expansion of nuclear power and the amount of additional coal consumption to be made up by imports.

In the maximum fossil fuel strategies, it is estimated that between 90-130GWe of nuclear power capacity will be required by 2000; in the maximum nuclear expansion (estimated at 292GWe by 2000), it is forecast that some 120-180 million tonnes of coal will need to be imported by the year 2000.

2.5 Environmental impact associated with energy production and consumption can be local, regional or potentially global in type and, although there is often scope for ameliorating impact, the cost of so doing can vary considerably. It should also be recognised that consumers will sometimes be faced with the choice between an energy development associated with a low probability of a high environmental damage and the alternative of certain relatively low impact.

2.6 Nuclear power devleopment has some physical impacts which are unavoidable, but the main concern is with events which, although having a low probability of occurring, could result in (a) death or serious health damage through sudden exposure to high level radiation and very toxic materials; (b) long term accidental release of stored high waste into the biosphere; (c) possible nuclear weapon proliferation; (d) terrorist threat of exploding a nuclear device manufactured from stolen fissile material; (e) the possible social consequences of providing effective safeguards against (c) and (d).

2.7 The risk of fatal radioactive release from a nuclear power station is very low, estimated at 1 in 10^6 to 1 in 10^8 per reactor year. Over the whole 1976-2000 period, the risk of such an occurrence is, therefore, determined at between 1 in 350 to 1 in 60,000 for the minimum and maximum nuclear strategies respectively, indicating the effect of time and rate of expansion on the risk factor.

2.8 Of more concern are the potential consequences of the nuclear fuel cycle. The enrichment process is generally considered acceptably safe. No reprocessing of irradiated oxide nuclear fuel has yet been carried out successfully on a full scale basis. Also, no certain satisfactory solution has yet been found to the final long-term storage of high-level radioactive wastes in geologically stable structures, although certain possibilities are now beginning to be examined. And although irradiated fuel after removal from the reactor can be safely stored in cooling ponds for several years, there is evidence to believe that the risks of this operation do increase with the passing of time.

Assuming radioactive waste is to be reprocessed, it is estimated that between 1-2 reprocessing plants would need to be built in the Community by the early 1990's in the minimum nuclear strategies; with maximum nuclear expansion alternatives, some 8 plants would be required by 2000. Under these strategies, the total accumulated high-level radioactive waste by 2000 would vary from 13,000 to 23,000 m^3.

2.9 On the whole, we concluded that, although the risk of nuclear proliferation cannot be discounted, under a well-supervised operation in the Community, the likelihood is low compared to other methods that could be available to non-nuclear nations bent on the means to acquire nuclear weapons. The terrorist risk is of greater concern, and is directly related to the number of reprocessing plants built. If these are limited, we do not believe that the size and operation of the necessary security force would infringe upon civil liberties of the Community in a significant way, athough there are undoubted implications for the employees at such plants.

2.10 The principal concerns with increased fossil fuel use are:

i. accidental local disturbance from expansion of coal production, particularly open-cast mining;

ii. accidental local and regional marine pollution from offshore oil production and transportation;

iii. a potential local, or even regional problem, of solid waste and sludge disposal from increased coal combustion;

iv. the local, regional and possible global impact of atmospheric pollution from fossil fuel combustion.

The impact of all four items mentioned can be ameliorated at a cost through pollution control measures, and improved operating procedures and standards. Of the four, potential increased atmospheric pollution represents the major cause for concern.

2.11 With no additional control measures, the potential increase in atmospheric pollutants under the alternative energy strategies is between 15-60% for SO_2 (of which the share from coal is 70-90%), 20-40% for NO_x, 30-60% for particulates. The local and regional impact is considerably influenced by other factors such as weather and topographical features. A recent OECD study has shown 20-60% of SO_2 and NO_x emitted from the EEC is exported. Imports of these pollutants would appear to offset this figure by about 5%.

Insufficient evidence currently exists to substantiate, or disprove, the possibility that carbon dioxide emissions could affect global or regional weather patterns.

SO_2 emissions from increased oil consumption can best be controlled by reducing the sulphur content of the fuel. Improved means for controlling sulphur emissions from coal combustion should be commercially available by the time an increase in coal combustion is likely. Depending on plant load factor, this could add 15-30% to the real delivered cost of energy from this source.

2.12 The impact of renewable energy sources is largely confined to local amenity disturbance involved. In some cases ,this will be considerable. Tidal power schemes will also affect, not necessarily adversely, local littoral and marine eco-systems. Offshore wave machines could well present a shipping hazard.

2.13 Expansion of electricity generating and transmission capacity carries a very marked impact in terms of local visual disturbance, increased requirement for cooling water (currently accounts for 35% of total Community fresh water abstractions), need for land (an additional 135-220 sites required by 2000 occupying some 11-17 x 10^3 hectares, and thermal pollution (100-275% increase). The need to acquire suitable new sites will represent the most significant impact.

3. Recommendations

Recommendations on policies to minimise the environmental impact associated with meeting future Community energy requirements are given on pages 33-36 of the report. These can be summarised as follows:

i. achievement of maximum feasible levels of energy conservation;

ii. giving consideration to the restriction of electricity consumption to essential uses;

iii. restricting the rate of expansion of nuclear power capacity so that:

a. a proper testing period for initial full scale reprocessing plant(s) can be undertaken;

b. to prevent the build-up of a large inventory of high-level radioactive waste before a satisfactory solution to safe long-term disposal has been found;

c. to allow time for the development of alternative methods for handling and disposing of irradiated fuel elements;

d. to ensure that the building of initial full-scale Fast Breeder Reactor plant(s) should be followed by a suitable test period, and that further development should only then take place if economically justified vis-a-vis other energy alternatives.

A high rate of nuclear expansion would also represent an inflexible energy policy and be susceptible to the consequences of public reaction to a nuclear accident;

iv. encouragement of indigenous fossil fuel production and coal imports;

v. further support for development and appraisal of renewable energy sources.

1. INTRODUCTION

1.1 In June 1977, ERL was asked by the Environment and Consumer Protection Service of the Commission of the European Communities to evaluate the potential environmental consequences of alternative energy strategies within the EEC up to the year 2000.

The primary objectives of the report were:-

i. to examine the effects on the environment resulting from the production and use of different forms of energy within the Community;

ii. to determine the scope for minimising environmental impact through the adoption of appropriate energy strategies, recognising the importance of other factors in formulating energy policies.

This is the final report. In December 1977, ERL submitted an interim report, and on 24th January a presentation of the summary and main conclusions of the final report was made at the time of the second session of the Commission's Nuclear Hearings in Brussels.

1.2 This report contains in Sections 2 and 3, the summary, conclusions and recommendations presented at the time of the Nuclear Hearings. These sections contain a few minor alterations made since that time.

1.3 Necessarily, all forms of energy production and use have an impact on the environment. The extent of that impact will depend first on the <u>quantity</u> of energy required and the <u>form</u> in which it is supplied and consumed; and secondly on the <u>measures</u> taken to reduce the impact.

In the report we have, therefore, examined:

- the future energy needs of the Community up to the year 2000 and the form in which the energy might be supplied, taking account of the projections of the CEC and Member States;
- the environmental and related impacts of the available options.

1.4 The main body of the report is, therefore, divided into two parts.

Part A contains the assessment of future Community energy needs under different sets of assumptions, and the forecasts of primary energy supply according to various alternative energy strategies.

Part B contains a description and analysis of the potential environmental consequences resulting from the production and consumption of different sources of energy, an assessment of the degree to which possible effects might be contained, and an evaluation of the overall impact on the Community under the alternative energy strategies identified in Part A.

1.5 It should be understood that in a wide-ranging study of this kind,a very large number of complex issues have had to be considered. Clearly, it would not be possible to treat fully in the text all the technical matters relevant to the issues being studied. However, we have examined in detail the latest available technical and economic data relating to different energy developments, assessment of their environmental impact and the means for controlling inherent pollutant and waste emissions; and, although the report is synoptic in approach, it does take a critical view of these various aspects.

1.6 In describing environmental effects, Part B of the report will distinguish between those which are global, regional and local in impact. It will be apparent that the degree of local impact can often be dependent on a number of specific site factors, and that by taking these factors into account, certain measures can be adopted which will ameliorate particular kinds of environmental damage. However, it is not the purpose of this report to describe how the effects of energy development can be reduced in this way, but to direct itself to identifying those energy policies whose adoption will minimise the overall environmental impact inherent in meeting the EEC's future energy needs.

1.7 To analyse the scale and type of additional energy supplies necessary to meet the future needs of the Community, certain assumptions have to be made. By describing these briefly - a fuller account is given in Sections 4 and 5 of the report - the approach taken by the consultants in deriving alternative energy strategies for the future will become apparent.

The EEC's future energy requirement is related to its level of economic growth*. For the purposes of the report we have made the following assumptions on GDP growth in the Community.

1976-1985 @ 4.2% per annum

1985-2000 @ 2.7% per annum

The first growth figure represents the combined weighted average GDP forecast for the Community as assumed in the 1976 national energy programmes (1). The growth assumption of 2.7% represents the average and middle order figure of economic growth of other recent long term energy growth forecasts for Western Europe made by the Workshop on Alternative Energy Strategies (2), and by the Energy Research Group at Cambridge (3), and by BP (4). These growth assumptions are also compatible with the assumption that the real price of energy will remain constant in the 1976-1985 period, and rise by 75%-100% in the 1985-2000 period.

---

* This relationship is critically examined in this study.

It may be that growth levels of this magnitude are not sustained. We have, therefore, also considered a low GDP assumption of 2.7% p.a. 1976-1985 and 1.8% p.a. 1985-2000. Growth rates could, of course, be higher than those first assumed; we have also considered a high growth scenario of 4.2% p.a. 1976-1985, and 3.5% p.a. 1985-2000.

As already stated, the principal objective is to examine the impact of alternative supply strategies. In theory these strategies can be adopted whatever the future level of economic growth although, as will be seen, some are more flexible than others. In selecting alternative strategies, the guiding principle has not so much been to choose those most likely or appropriate for energy policy makers, but more to include a range of strategies which pursue possible courses of action to their feasible maximum (or minimum) so that the resultant environmental consequences will be highlighted.

We have developed two sets of energy strategies 'A' and 'B, defined by the degree of energy conservation adopted.

ALTERNATIVE ENERGY STRATEGIES	
Price induced conservation strategies	Maximum conservation strategies
1A Maximum nuclear	1B High nuclear
2A Maximum fossil fuels	2B High fossil fuels
3A Maximum 'renewables'/high nuclear	3B Maximum 'renewables'/high nuclear
4A Maximum 'renewables'/high fossil fuels	4B Maximum 'renewables'/high fossil fuels

The maximum level of nuclear expansion in 1A is determined by the estimated feasible level of power station construction in the Community. In the case of fossil fuels (2A), the maximum is determined by the limits imposed by forecast indigenous production and the assumed maximum level of imports available. In both sets of 3 and 4 strategies, renewable sources of energy (including Combined Heat and Power systems and energy pumps) are developed to what are considered their feasible maximum levels in the 1976-2000 period; in strategies 3A and 3B, the balance of primary energy supply is made up by nuclear power, with only limited fossil fuel expansion; in strategies 4A and 4B, the balance of primary energy supply is made up by fossil fuels, with only limited expansion of nuclear power capacity beyond that currently being built, under construction, ordered or firmly planned.

The point needs to be made that in assessing the potential contribution of different sources of energy, cost considerations did not influence the levels adopted within strategies. The aim is to determine the maximum feasible levels of expansion in the time period, given the availability of resources, manpower and development lead times. This is not to deny that costs will play a major role in shaping future energy strategies but, as will be argued in the conclusion, there is considerable uncertainty about the future costs of different forms of energy, including nuclear power, and their relative economic attractiveness could have shifted radically by the 1990's.

These considerations also apply to energy conservation. In all strategies adopted, a certain amount of energy saving will result from consumer responses to the movement of energy prices in the 1973-2000 period. The maximum conservation 'B' strategies imply that in addition to price induced conservation, deliberate policies are adopted by governments through subsidies, fiscal incentives, legislation, R and D support, etc., to achieve markedly higher levels of energy savings without introducing significant alterations in the way of life of the Community.

1.8 The approach taken in Part B in evaluating the impacts under the alternative strategies was to generate the associated levels of pollutant emissions, waste and resource requirements assuming that current standards of unit emission levels remain constant, thus identifying the potential increase in impact; then discussing this increase in the light of additional means for control that might be available, taking into account the costs of introducing such controls.

1.9 We have prepared the conclusions to the study in the form of answers to a series of questions. These can be grouped under the following general headings:

1. What are the future energy requirements of the EEC and what factors affect these requirements?

2. How can future energy demands be met?

3. What are the impacts of alternative energy strategies, and to what extent can they be ameliorated?

2. CONCLUSIONS

2.1 1976-2000 Community energy demand

- TO WHAT EXTENT DOES FUTURE ENERGY DEMAND DEPEND UPON THE RATE OF ECONOMIC GROWTH?

The relationship between the rate of growth of primary energy demand and that of GDP, is normally expressed as an energy coefficient*. It is extremely complex and is the macro-economic result of very many energy consumption/consumer behaviour patterns in different sectors of the economy. These in turn are influenced by many factors including the price of energy, level and type of industrial output, applied technology, overall standard of living, demography etc. As such, the energy coefficient is liable to considerable change over time, as is demonstrated by the sharp rise that occurred between the two ten year periods following the Second World War (approximately 0.75 to 1.2). It is forecast that a considerable reduction is likely over the 1976-2000 period, compared to the 1960-1974 figure of 1.0.

	Forecast Energy Coefficient 1976-2000	
	Unrestricted growth ≁	Max. conservation
Base case	0.68-0.72	0.54-0.55
High economic growth	0.62-0.65	0.50-0.54
Low " "	0.72-0.74	0.58-0.64

The following points can be made.

i. Although a higher rate of economic growth will lead to a larger final total energy consumption for the Community, the increase is very much less marked in the transport and domestic/commercial sectors than in the industrial sector. This phenomenon will give rise to improved (lower) energy coefficients at higher rates of economic growth.

ii. This unlinearity referred to in (i) arises because, for any given level of population in the Community, there is likely to be a flattening off in demand for automobiles, improved domestic heating, and basic consumer durables as higher standards of living lead towards saturation of these markets. Obviously, saturation levels are never actually reached and considerable uncertainty surrounds them. Nevertheless,

---

* The ratio of rate of annual increase in primary energy demand to rate of annual increase in Gross Domestic Product.

≁ Takes into account price only induced conservation.

estimates can be made of these parameters, and it is to be expected that by reaching a higher standard of living in the Community more quickly, it will follow that the flattening off in transport and domestic energy demand curves will occur earlier.

iii. Because the flattening of domestic and transport sector energy demand curves is unlikely to occur much before 1990, and the rise in the real price in energy is forecast to occur after 1985, the most marked lowering of energy coefficients is likely to occur towards the end of the period of study.

- DOES FUTURE PRIMARY ENERGY DEMAND DEPEND UPON THE WAY IN WHICH ENERGY IS CONSUMED?

Consumers require energy for a number of different purposes, e.g. space heating, water heating, mechanical drive, process heat, electrolytic ore reduction, etc. The amount of energy actually required for the function in question is referred to as "useful heat". However, as indicated in Figure 1 on page 9, there is an enormous difference between the "heat supplied" and the "useful heat" consumed. In 1975, overall useful heat consumed* in the EEC was 325 million tonnes oil equivalent⁄, whereas "heat supplied" to final consumers was 712 million tonnes oil equivalent. The efficiency of use varies according to use. Some of this heat loss is inherent to the physical process taking place and can never be recovered but, as will be argued later, a considerable proportion can be conserved. For certain uses, energy has to be supplied in a particular form, e.g. electricity for lighting, certain kinds of mechanical drive, etc. However, in most applications one form of energy can be substituted by another. And because some forms of energy are inherently more efficient for certain purposes, energy policies should be geared to optimising the use of energy resources relative to the availability, e.g. natural gas can be burnt as a domestic fuel more efficiently than oil or coal.

- DOES THE FORM IN WHICH ENERGY IS SUPPLIED INFLUENCE THE OVERALL LEVEL OF FUTURE ENERGY REQUIREMENTS?

Energy is not always consumed in the same form as the primary energy source, e.g. crude oil, coal, solar, hydro-power, etc., but is converted and delivered as secondary energy. This conversion process itself involves the consumption and loss of primary energy. As illustrated in Figure 1, this loss as a % of primary energy input is relatively small in oil refineries (8%), coalmines (3%), more marked in coke ovens (22%) and synthetic coal gas plants (about 35%), but

---

* Includes non-energy oil consumption.

⁄ 1 tonne oil equivalent = 1.43 tonnes coal equivalent.

when primary energy is converted to electricity in thermal power stations, the minimum attainable heat loss is about 59% and average loss today is nearer to 68%. Very little of this waste heat is utilised at the moment.

It therefore follows that when electricity, thermally generated from fossil fuel is used as a source of space or water heating, three to five times as much primary energy source is being wasted as when the fossil fuel is burnt directly. This estimate takes account of transmission losses of electricity (5%-9%), offset by its virtually 100% efficiency of consumption.

- WHY IS THE FUTURE DEMAND FOR ELECTRICITY OF PARTICULAR CONCERN?

The reasons for minimising electricity consumption derive not only from the enormous waste of primary energy sources inherent in thermal generation but, also, as will be argued later, from the scale and type of environmental impact and the potentially serious social and political consequences associated with its generation, and also because of the high financial risks associated with the expansion of large capacity generating units. Therefore, the question that should be asked by energy policy makers is "what are the essential future requirements of electricity and how can these demands be most desirably met?", and not "what new markets can be found for electricity?".

- WHAT IS THE MINIMUM DEMAND FOR ELECTRICITY?

In the industrial sector (approximately 42% of total consumption), where electricity's essential uses are in providing lighting, mechanical drive, direct reduction of metal ore and electrolysis, consumption growth is likely to continue to rise approximately in line with industrial/economic growth. However, from the late 1980s onwards, there may be a shift of the expansion of high electricity consuming industries, such as aluminium smelting/petrochemicals, to developing countries where there is cheaper electricity and ready sources of chemical feedstock.

In the domestic/commercial/miscellaneous sectors, heating use currently accounts for about 15% of total Community electricity production. Growth in use of lighting and electrical appliances is currently growing at about twice that of economic growth. This pattern is likely to continue until the middle 1980s, but a marked slowing down is expected as some markets begin to approach saturation in the last 15 years of the century (even allowing for new markets, e.g. expansion of telecommunications).

On the base case economic growth assumptions, the minimum electricity consumption growth is forecast at 4.6%-4.9% per annum 1976-1985, falling to 2.2%-2.4% per annum 1985-2000. Even this growth implies a 2000 electricity consumption about 2.1 times the 1976 level. The higher nuclear growth strategy would increase this by about another 30%.

The minimum electricity consumption in 2000 will not vary in direct proportion to economic growth in the intervening period. It is estimated that the low GDP growth scenario will result in a reduction of the base case 2000 postulated level of consumption by 14%-20%. Raising GDP growth in 1985-2000 from 2.7% to 3.5% per annum is estimated to raise minimum 2000 electricity consumption by about 10%-13%.

- TO WHAT EXTENT CAN CONSERVATION MEASURES RESTRICT FUTURE ENERGY DEMANDS?

Energy conservation will take place for two reasons:

i. because of consumer response to energy price increases that have already taken place in the last five years, and those which are forecast to take place in the next twenty years or so;

ii. because of government action which will encourage or require consumers and manufacturers to reduce their own delivered energy consumption or the energy requirements of their products respectively; in addition, ways can be found for utilising heat that is currently wasted by secondary energy producers and by industrial consumers; governments can improve energy conservation through a number of different measures including subsidies, fiscal incentives, legal constraints, R and D support and planning regulations.

In this study, both types of conservation are taken into account, type (i) in all energy strategies, type (ii) only in the maximum conservation 'B' strategies.

The scope for maximum conservation achieved through vigorous policies varies enormously according to the consumer. The estimates of the amount by which categories of consumers can reduce their energy consumption is summarised below. As mentioned, they do not envisage introduction of laws which significantly alter life styles in the Community.

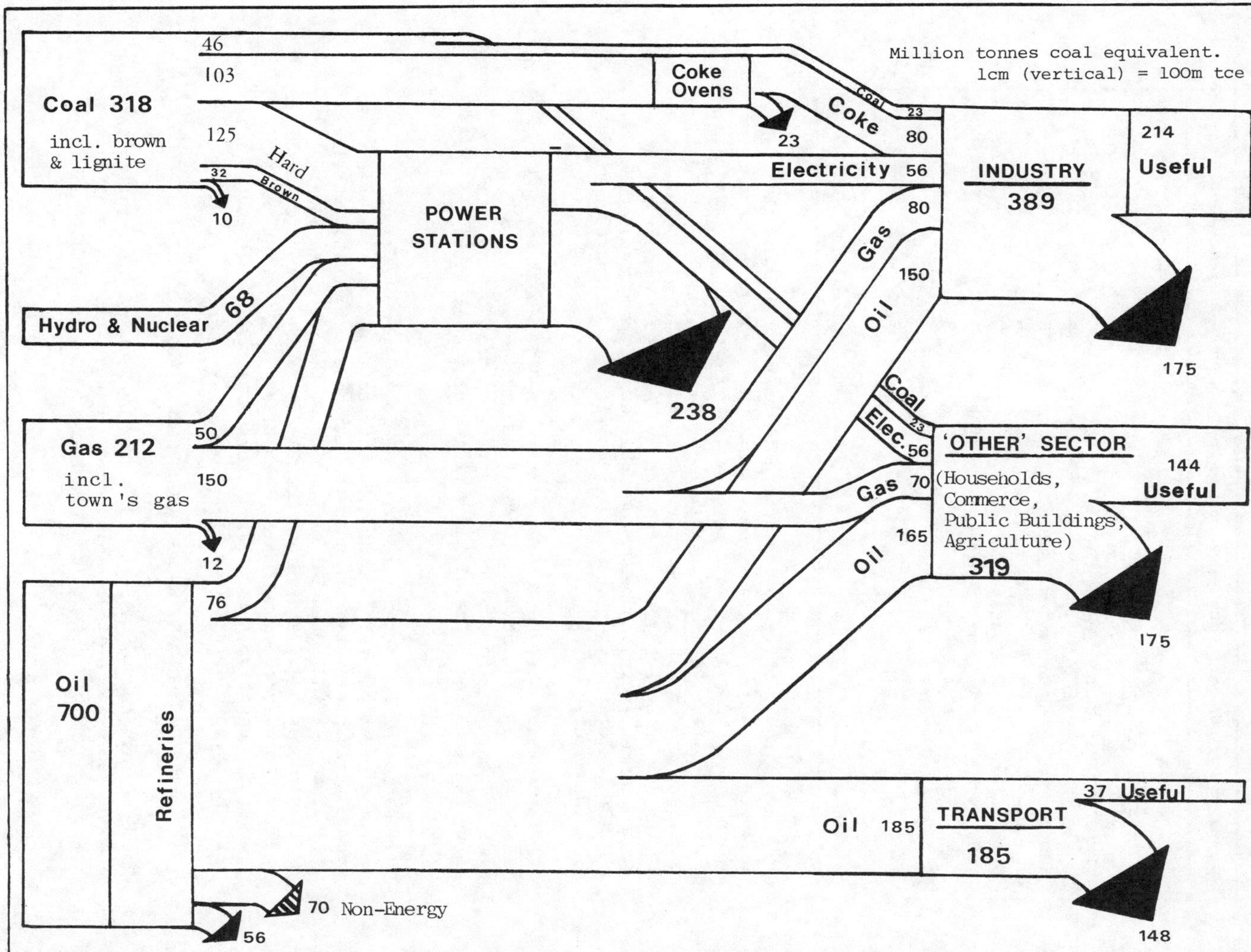

Figure 1. ENERGY FLOWSHEET FOR THE EEC

Table 1

ESTIMATED SCOPE FOR 'MAXIMUM' CONSUMER ENERGY CONSERVATION

Consumer Category/Energy Use		% by 1985*	% by 2000*
Transport	- road	6	13-16
	- air	7	15
Domestic	- space/water heating	16	37-42
	- electrical appliances	5	15
Commercial	- space/water heating	15	30
Industry	- iron and steel	7	20
	- aluminium	8	18
	- chemical	10	20
	- other	14	25

* From 1975 levels

It will be noted there is less scope for conservation in the high energy consuming iron and steel, aluminium and chemical industries, which account for about half total industry energy consumption, since these industries have historically had an incentive to be more efficient in their use of energy. The figures clearly show that it is the domestic sector which apparently deserves the most attention of government energy conservation policies, although successful realisation of these potential savings will be hardest to achieve in this sector.

The increasing energy price forecast in the 1985-2000 period is likely to lead to a structural shift towards expansion in lower energy consumer industries. The effect of this by the year 2000 is unlikely to reduce total industrial energy consumption by more than about 5%.

Utilisation of waste heat from thermal power stations (combined heat and power plants) through district heating schemes is restricted to those power stations near to high density housing. It is estimated that the potential of CHP in the Community is 100-125 million tonnes o.e., equivalent to 30%-35% of the domestic heating market. However, because of the lead times involves (up to 15 years per project), it is unlikely that more than one third of this waste heat could be utilised by the year 2000.

- WHAT ARE FUTURE PRIMARY ENERGY REQUIREMENTS FOR THE EEC?

These are assessed on a detailed sector-by-sector basis, taking into account the factors already discussed. Future demands relating to the base case economic assumptions taken are forecast as follows:

Table 2

FORECAST 1976-2000 EEC PRIMARY ENERGY CONSUMPTION					
Strategy	1976 10^6 tonnes p.a.	1985 10^6 tonnes o.e.	1976-1985 p.a. growth %	2000 10^6 tonnes o.e.	1985-2000 p.a. growth %
1A	969	1298	3.2	1736	1.9
2A	969	1298	3.2	1685	1.8
3A	969	1298	3.2	1666	1.7
4A	969	1298	3.2	1660	1.7
1B	969	1254	3.0	1529	1.3
2B	969	1254	3.0	1520	1.3
3B	969	1254	3.0	1505	1.2
4B	969	1254	3.0	1493	1.2

A detailed assessment of the primary energy requirements generated at a low economic growth rate (2.7% and 1.8% per annum) over the two periods) yielded a 2000 primary energy consumption some 16%-19% below the equivalent base case estimates above. We estimate that a high economic growth rate throughout the 1985-2000 period would give rise to a 2000 energy consumption about 5%-10% above the figures shown for the equivalent strategies in Table 2.

The following points can be made:

- primary energy consumption is likely to increase by 55%-79% from 1976 to 2000, depending upon the energy strategy adopted, assuming a middle order rate of economic growth (average 3.2% per annum throughout the period);

- maximum variations in these figures at low and high economic growth rates are estimated to be -15% and +10%, i.e. giving a maximum range for the increase of 40%-89%;

- variations in the long term primary energy consumption requirements are just as likely to depend on the energy strategies adopted than on uncertainty about future economic growth;

- in all cases a slowing down in the growth rate of energy consumption is expected after 1985.

- HOW DO THESE PRIMARY ENERGY CONSUMPTION FORECASTS COMPARE WITH OTHERS THAT HAVE BEEN MADE?

Exact comparison cannot be made since no other published energy forecasts for the EEC have been made for the year 2000. However, the MIT sponsored Workshop on Alternative Strategies (3), the Cambridge Energy Research Group for the 1977 World Energy Conference (2) and Mr R. Belgrave of BP (4) have made estimates of primary energy demand for the year 2000 for Western Europe. If it is assumed that 2000 EEC energy consumption will be 77% of total Western Europe energy demand (cf 80% in 1976), then the following comparative points can be made:

i. ERL's 1976-2000 average forecast energy growth (average 1.5% -2.7%) is similar to WAES (1.6%-2.6%) and World Energy Conference (1.9%-2.2%), but significantly lower than that of Belgrave (2.7%-3.6%);

ii. however, ERL examined a wider possible variation in economic growth, and our forecasts imply a greater reduction in energy coefficient, i.e. greater potential energy savings/more efficient use of primary energy, (0.56-0.72) than WAES (0.63-0.79), and the World Energy Conference (0.61-0.79). Belgrave made no explicit assumptions about economic growth.

2.2 Energy Supply

- TO WHAT EXTENT CAN THE EEC RELY UPON HYDROCARBON FUELS FOR ITS FUTURE ENERGY SUPPLY?

In 1976, 72% of the Member States' total energy supplies came from oil and natural gas (542 and 156 million tonnes o.e. respectively).

Indigenous production

Of the 1976 oil consumption, only 22 million tonnes was produced indigenously. With the expansion of North Sea production and exploration proceeding in other areas, the potential for future indigenous hydrocarbon production is given below. These estimates are based upon a general consensus view of the forecast for 2000, and are liable to variations of around ±15%.

Table 4

FORECAST EEC HYDROCARBON PRODUCTION*

Million tonnes o.e.	1976	1985	2000
Natural Gas	144	150	105
Crude oil	22	140	110
	166	290	215

* Includes natural gas liquids

The decline in natural gas production stems largely from the depletion of the Dutch Groningen field. North Sea oil and gas production will probably peak in the middle to late 1980s and decline slowly thereafter.

Imports

Natural gas from the Norwegian sector of the North Sea, pipeline imports from the USSR (Iran) and Algeria and seaborne Liquified Natural Gas from OPEC countries will form an increasingly large share of total natural gas supplies. Potential OPEC exports of natural gas, although costly to liquefy and transport, are very large and the EEC is geographically well placed to compete for them.

The potential for increased crude oil imports is very much more constrained. In 1976, OPEC production was about 32 million barrels/day of which the EEC share was about 1/3. Given the current known and speculative reserves and Saudi Arabia's lack of financial incentive to maximise its production, it is unlikely that exports from this source will exceed 40-45 million barrels in the 1990s. Although production from new areas of the world will come into the market, so too will oil demand increase in developing countries, which currently import only small quantities of crude oil. With these points in mind, the potential for future hydrocarbon imports is set out below:

Table 5

FORECAST POTENTIAL HYDROCARBON IMPORTS

Million tonnes o.e.	1976	1985	2000
Natural gas	12	79	100-120
Crude oil	520	675-725	600-675
	532	754-794	700-795

- WHAT IS THE POTENTIAL OF INDIGENOUS COAL PRODUCTION?

The UK has very large reserves of coal, which will allow expansion of new deep-mined production to more than offset the run-down of old pits. West Germany will also be able to expand its open-cast production, but elsewhere in the Community coal production is expected to decline.

Table 6

FORECAST INDIGENOUS COAL PRODUCTION			
Million tonnes hard coal equivalent	1976	1985	2000
Deep-mined	236	207	237
Open cast	48	56	61
	284	263	298

It can be seen that the overall increase is small.

- WHAT IS THE POTENTIAL FOR RENEWABLE SOURCES OF ENERGY?

The combined contribution of indigenous coal production and total hydrocarbon supplies is insufficient (1123-1218 million tonnes o.e.) to meet forecast EEC energy demand in 2000. Can, therefore, the difference be made up from 'renewable' sources of energy? The outlook for each source is discussed individually in the main report. By far the most important potential contribution could come from the installation of heating panels in buildings, the technology of which is already well developed. Allowing for the orientation and type of existing housing stock, the amount of new building that is likely to take place by 2000 and the latitude of settlements in the Community, it is estimated that the maximum contribution by the year 2000 of solar energy in providing water and space heating in domestic/commercial buildings is 50 million tonnes o.e.; and as low temperature process heat for industry, the estimate is 17 million tonnes o.e. Heat pumps utilising residual heat in the environment are already in use and could also contribute some 5-10 million tonnes p.e. (net) to space heating.

There is much greater uncertainty associated with other forms of renewable energy. The technology of wave power is very much in its infancy. For large scale wind turbines, solar voltaic cells and refuse-derived fuel, the technology is at a more advanced stage but the economics of utilisation or their reliability are still much in question. Tidal power schemes would require very detailed site engineering studies and environmental impact assessment and large projects would take 10-20 years to complete. The tapping of geothermal power in Italy and in France also requires much more knowledge of geological structures and several exploration wells would have to be drilled. Except for refuse-derived fuel (the collection and transportation of biomass would seem too high for it to make much contribution this century), all those other 'renewable' sources of energy would be utilised as primary sources of electricity. Wave, wind and tide power are either uncertain or periodic in their availability. Therefore, unless some associated means of cheap energy storage is developed, their contribution should be primarily seen in terms of savings of traditional energy forms, rather than as substitute electricity generating capacity. Wave power has much the greatest potential contribution to make, but on the other hand is probably the most expensive and is at the earliest stage of development.

Their total maximum contribution by 2000, taking a reasonably optimistic view of the time required for development and installation is assessed as 45-50 million tonnes o.e.

Total maximum renewable sources of energy by 2000 is therefore estimated at 114 million tonnes o.e. Waste heat and heat pumps contributions are taken into account under energy conservation.

- WHAT IS THE POTENTIAL CONTRIBUTION OF NUCLEAR POWER?

An assessment of the capacity and capability of the nuclear power industries of Member States and the likely availability of skilled workers would not permit more than 293 GWE of nuclear power capacity to be installed by the year 2000, equivalent to 400 million tonnes o.e. Such a programme would also require a very substantial expansion in nuclear fuel enrichment and fabrication plants, with an estimated maximum of 10 by 2000. The need for fuel reprocessing plants is arguable with a low level of thermal nuclear capacity, but with the maximum nuclear expansion programmes, a total of 6-8 reprocessing plants would seem unavoidable. It is worth pointing out that no full-scale reprocessing plant to handle irradiated fuel from Advanced Gas Cooled or Light Water reactor power stations has yet been operated successfully. The two U.S.A. plants are current shut down, and France's La Hague plant currently has a capacity of only 400 tonnes/year.

From what is known of proven or likely ore reserves, it would seem likely that sufficient uranium fuel can be obtained by Member States to meet a nuclear programme of this scale. This raises the question of Fast Breeder Reactors. Even assuming a successful five year test period for the French Super-Phenix FBR station (due to be commissioned in 1983) and also that the UK does decide to go ahead with a full-scale test plant CFR-1, it is unlikely that more than 3 or 4 FBR stations would be installed in the Community by the year 2000.

- WHAT OTHER ALTERNATIVE IS THERE TO NUCLEAR POWER TO ENSURE PRIMARY ENERGY REQUIREMENTS IN 2000 ARE MET?

It can be seen that the contribution of 'renewables', even assuming maximum conservation policies, is unlikey to be sufficient to ensure that total primary energy demands by 2000 are met. Besides nuclear power expansion, the only other potential alternative to fill the so-called 'energy gap' is, in the absence of unlimited foreign oil supplies, to import greatly increased quantities of coal.

Currently, the Community imports about 40 million tonnes of hard coal/year, mainly for coking purposes. The potential for increased coal exports from South Africa, Australia, Poland and, to a lesser extent, from the United States is very substantial. Indeed coal is eventually likely to take over from oil in supplying the marginal unit of the world's energy requirements.

Taking into account Japan's long term coal input requirements, the EEC might expect to be able to import 200-275 million tonnes of hard coal by the year 2000, equivalent to 140-192 million tonnes of oil. Again it should be pointed out that, although the reserves are sufficient to support such an export programme, there is considerable uncertainty attached to the rate at which respective producer governments permit the developments to proceed, and to whether the necessary mining and supporting infrastructure investment will be forthcoming.

- WHICH IS THE LEAST COSTLY COURSE OF ACTION IN ENSURING FUTURE ENERGY REQUIREMENTS ARE MET?

Comparisons of costs between alternative means of supplying additional quantities of energy are very difficult and may well be misleading for the following reasons:

i. available data has not always been produced on a similar basis nor at the same time;

ii. knowledge is incomplete in many areas;

iii. there is considerable uncertainty on relative rates of inflation of different materials and labour;

iv. future technology development could significantly alter the relative costs of one indigenous source of energy vis-a-vis others.

The capital costs of nuclear plants and most alternative energy producing/processing plant have recently risen very steeply as a result of the high inflation that has been associated with plant construction costs. For nuclear costs, even more striking has been the upward reassessment of the likely real costs arising from the fuel cycle, particularly that of reprocessing. Nuclear system capitalised costs are now put at $1345-$1454/kW (1976 prices) including interest charges. Final waste disposal and plant decommissioning costs could add a further 5% or so to this figure.

Electricity generated from PWR nuclear power stations has been estimated as costing 2.7-3.2 cents/kWh at 60% load factor (adjusted to 1976 prices). These costs compare with those that have been quoted for other 'renewable' primary electricity sources, e.g. the Severn tidal power barrage 3.8-4.6 cents/kWh, wave power system (two types) 6-12 cents/kWh, geothermal power 1.0-4.5 cents/kWh, wind power 2.75-4.1 cents/kWh. We would remark that the nuclear cost estimates have been prepared on a much more detailed assessment of the various material and labour components than the 'renewable' energy sources and could perhaps be considered more reliable. However, no operational experience has yet been gained with irradiated fuel reprocessing, nor are the final costs of disposal included. Also, it is possible that the nuclear industry will have to reduce low level radiation exposure levels to plant works, which could again add to nuclear generating costs.

Lowering the interest rate used in discounting future costs would also favour those projects with very long load times, e.g. combined heat and power systems and tidal power schemes. On this evidence, we would conclude that nuclear power does not hold any overriding economic advantages over these alternatives.

Future costs of electricity from coal stations and estimates of Synthetic natural Gas are very difficult to make with confidence because of uncertainty over the coal price. However, both would have a small cost advantage over nuclear power, assuming current oil prices equivalent for coal at 60%, and 65% load factors for power stations and SNG plant respectively.

The point can also be made that the potential economic risk with large-scale projects having long development load times is much greater than that associated with small-scale projects. The risk arises not only from the uncertainty involved with forecasting long term energy demands correctly, but from the imperfect knowledge of future costs and energy prices. In this respect the introduction of energy conservation measures and solar panel installation would have a clear advantage over the building of large electricity generating capacity. It should also be mentioned that a recent UK estimate of the average cost of 15% energy savings to industry was equivalent to about 1 cent/kWh (amortising capital cost of investment over 15 years @ 10%). This figure is significantly below alternative energy supply costs.

- WHAT THEREFORE ARE THE PRIMARY ENERGY FORECASTS IN THE ALTERNATIVE SUPPLY STRATEGIES?

The base case forecasts are shown in Table 7 below.

Table 7

FORECAST EEC PRIMARY ENERGY BALANCES - 1976-1985 AND 2000

Million tonnes o.e.	1976	1985			2000							
		1A/ 3A	2A/ 4A	B Strategies	1A	2A	3A	4A	1B	2B	3B	4B
Solid fuels	220	236	242	235	320	400	325	381	280	360	275	370
Oil	542	658	668	631	715	742	641	740	630	680	557	637
Natural Gas	156	229	229	229	215	215	215	215	215	215	215	215
Hydro	20	34	34	34	35	35	35	35	35	35	35	35
Nuclear	20	140	124	124	400	240	346	185	316	177	309	124
Renewables	1	1	1	1	53	53	114	114	53	53	114	114
TOTAL	969	1298	1298	1254	1736	1685	1676	1670	1529	1520	1505	1493

The following points are made:

- there is considerable variation in the need for additional nuclear power capacity, beyond that already installed, under construction or ordered (about 70 GWe capacity) according to the degree to which other supply/conservation options are exercised;

- there would appear to be a need to consume at least 25%-45% more coal over 1976 levels, even with a policy of high nuclear power expansion;

- a lower or higher economic growth rate would affect most significantly the requirement for nuclear power and/or imported coal.

- WHAT ARE THE STRATEGIC IMPLICATIONS OF THE ALTERNATIVE SUPPLY STRATEGIES?

By opting for a high nuclear strategy the requirement for imported energy to the Community is 42%-47% (depending on the level of conservation and 'renewables' contribution) compared to 54%-55% in the high fossil fuel strategies. This compares with the 1976 imported energy contribution of 57%.

2.3 Environmental Impact

- WHAT ARE THE PRINCIPAL FORMS OF ENVIRONMENTAL IMPACT?

The production, transport and use of the available types of energy give rise to many different forms of environmental damage; the extent of such damage tends to be determined at least in part by local factors. In considering the impact of the alternative strategies we are concerned principally with the types of damage inherently associated with the supply options. In this context, the following are relevant.

- Damage from energy production, transport and use may have local, regional or global implications.

- The principal form of local damage that is difficult to ameliorate is the effect on amenity arising from facilities for the extraction and processing of fossil fuels for generating and transmitting electricity. Disposal of solid wastes can present significant local impact and land set aside for the disposal of nuclear wastes must also be classed as a serious disamenity.

- Atmospheric pollution caused principally by fossil fuel use has both a local and regional and possibly global impact. The effects on health, vegetation and materials of sulphur dioxide, NO_x and particulates are well documented. It is now generally accepted that SO_2, NO_x, and small particles may be transported large distances. Recent OECD investigations show the EEC to be net exporters of SO_2 to other countries.

- The accidental release of high radioactive material during the burning of nuclear fuel or during the operation of the fuel cycle, including its final disposal would cause serious local or regional damage of either a short or long term nature. Low level radioactive release also occurs in effluent emitted from nuclear plants, and from waste disposal. Also associated with nuclear power development is the possible threat of nuclear proliferation and terrorist activity.

These in our view are the three principal forms of impact: disturbance at the local level, effluent in the atmosphere and possible releases of radioactivity. There are, of course, many others: thermal pollution from discharge of cooling water, noise, secondary impacts associated with increased water needs, possible weather disturbances from CO_2, particulate emissions, unavoidable and accidental effects of offshore oil developments, etc.

It is important to stress that these different forms of impact cannot be directly compared. There are two important components to any environmental impact evaluation: the impact itself and the way this impact (or risk of impact now or in the future) is perceived by the community or society at large.

- TO WHAT EXTENT CAN THE IMPACT BE AVOIDED?

The ability to avoid or restrict environmental damage associated with the supply, transport and consumption of energy is dependent both upon the type of impact and the form of the primary or secondary energy source in question. Similarly, the cost of ameliorating impact can vary enormously, which has of course a direct bearing on the feasibility of controlling pollution. Available and prospective measures for control will be discussed below. However, it is inevitable that, although there are several energy policies available which will always tend to reduce overall environmental damage, some additional impact will result from whatever energy strategy is adopted. Minimising the environmental consequences of the EEC's future energy development therefore becomes a matter of choice between alternative impacts, or an appropriate balancing act.

- WHAT ARE THE PRINCIPAL ENVIRONMENTAL IMPACT OPTIONS FACING POLICY MAKERS?

It follows from the conclusions on available supply alternatives that the Community, in meeting its forecast energy requirements in the 1990's, will be facing:

i. the low probability but potentially very serious risks associated with an expanded nuclear programme, and also the unavoidable amenity impact of this alternative; or

ii. the increased atmospheric pollution inherent in maximising fossil fuel consumption particularly coal; this alternative could also lead to solid waste disposal problems and some inevitable local amenity impact.

In reality, it is probable that no supply strategy is likely to exclude one of the two alternatives entirely. Nevertheless the option exists to minimise one or the other of the two types of impact, or decide upon a certain balance.

Conclusions on the type and degree of impact associated with these two options in the alternative energy strategies, will now be discussed, together with the scope for, and cost of, their control.

- WHAT ARE THE POTENTIAL IMPACTS OF EXPANDING NUCLEAR POWER CAPACITY?

Nuclear power has associated with its development:

i. some impacts which are unavoidable;

ii. events which only have a low probability of occurring but which, if they occurred, could result in any of the following:

   a. death or serious health damage through exposure to high level radiation and very toxic materials;

   b. possible nuclear weapon proliferation;

   c. terrorist threat of exploding a nuclear device, manufactured from fissile material stolen from the nuclear fuel cycle;

   d. the possible social consequences of providing effective safeguards against (b) and (c).

   e. long term accidental release of stored high level waste into the biosphere.

Implications of the nuclear fuel cycle

Unless supplied directly with enriched fuel from outside the Community, all the strategic alternatives would require the installation of uranium fuel enrichment and fabrication plants. (The amount of uranium ore likely to be mined in the Community is a small proportion of the requirements of even the minimum nuclear strategy, and so imported fuel arrives in the form of 'yellow cake' U_3O_8). The operation of these plants carries certain potential pollution threats because of the use of a dangerous chemical (uranium hexafluoride), which has to be carefully guarded against. The release of radioactivity is extremely small from these plants (but increases somewhat if recycled uranium or plutonium are included). In principle an accident could occur through the unintentional creation of criticality, which in turn might lead to a fracture of the containment system and a release of radioactivity. This is guarded against through the design of the plant and storage, and the exclusion of water from sensitive areas. In the minimum nuclear expansion strategies (4A, 4B) there would probably be no need to build more than 2-3 enrichment plants and a similar number of fuel fabrication plants, i.e. current facilities and those under construction sufficient enriched fuel for the maximum nuclear power strategy by 2000.

The situation with regard to reprocessing plant is not simple, as those who followed the 1977 UK Planning Enquiry on Windscale would acknowledge. Irradiated fuel from thermal nuclear power stations can be indefinitely stored, initially in nuclear ponds, while it cools down, and then by disposing of the waste in some acceptable fashion. The purpose of reprocessing is:

a. to recover reusable U-235 fuel, and plutonium - the principal fuel for Fast Breeder Reactors;

b. to separate the radioactive waste into low, middle and high activity streams, so as to reduce the volume of high-level nuclear waste, though the reprocessing operation itself also introduces additional problems of disposal.

Without irradiated nuclear fuel reprocessing, there is no separation of plutonium, the most suitable fissile material for the manufacture of a nuclear bomb. In the minimum nuclear expansion strategies 4B and 4A (90 Gwe and 119 GWe installed capacity in 2000), 2-3 reprocessing plants would be required if all waste were to be reprocessed. However, there would be no need for more than 1-2 of such plants to be built in the Community and be successfully operating before 1990 in these minimum strategies, assuming irradiated fuel were treated on a Community basis. The maximum nuclear expansion strategy would require around 6 reprocessing plants to be in operation by 2000, if all nuclear waste were to be reprocessed. Even if the maximum feasible amount of unprocessed irradiated fuel were stored, it seems unlikely that it would be possible to operate the maximum nuclear strategy effectively without 3 reprocessing plants being in operation by 1990 and 4 by the year 2000. Such a programme would hardly allow time for an effective test period for the initial oxide nuclear fuel reprocessing plants in the 1980's, and for the security implications of the complete nuclear fuel cycle to be fully explored.

Nuclear waste *

The nuclear waste generated in the alternative energy strategies is shown in Table 8.

Table 8

WASTE (1) FROM NUCLEAR POWER GENERATED BY 2000								
Strategy	1A	2A	3A	4A	1B	2B	3B	4B
Installed nuclear capacity GWe	292	175	253	131	227	129	222	90
Total depleted uranium 10^3 tonnes	932	685	855	607	794	600	794	548
Inactive wastes 10^3 m^3	419	305	385	273	365	257	357	223
Low level wastes 10^3 m^3	1000	800	900	700	700	800	700	600
Cladding wastes 10^3 m^3	12.0	8.8	11.0	8.0	9.4	7.7	9.4	6.9
High level wastes 10^3 m^3	23	17	21	16	20	15	20	13
Recycled U(2) 10^3 tonnes	171	104	157	111	149	105	146	91
Recycled Pu. 10^3 tonnes	0.95	0.70	0.87	0.63	0.75	0.61	0.75	0.55

(1) Assumes all waste is reprocessed.

(2) Contains about 0.8% U_{235}.

While the difference in annual output between the high and the low nuclear strategies is fourfold, the difference between the accumulated waste by the year 2000 is only about twofold. Even so, the total accumulated nuclear waste in the minimum nuclear strategy (which assumes no further ordering of nuclear capacity beyond that already firmly planned) is over twenty times that which had been produced by the beginning of 1978.

The very low levels of radioactive waste emitted from nuclear power stations in liquid and gaseous form during normal operation are not considered a problem, although if irradiated fuel is stored at power stations for too long, the potential for increased radioactive liquid effluent increases. The amount of radioactive low-level liquid waste from a reprocessing plant would, however, need careful monitoring. Low level solid radioactive wastes are buried, whereas intermediate level solid waste needs shielding in drums before burial at a suitable site. Some other intermediate activity waste and the radioactive fuel cladding are stored in silos at sites; in Germany, they are eventually embedded in concrete/bitumen and stored in a suitable site. Within certain limits, intermediate activity in wastes can be dumped in the ocean.

* A more detailed account of this subject is given in Sections 8.2.4, 8.2.5 and 12.1.3.

The main problems with respect to nuclear waste reside with the high activity liquid wastes. These contain fission products and also actinides (including plutonium) which are extremely toxic, and many have very long decay half-lives*. They must be kept entirely secure from the environment, some for thousands of years. Current methods of disposal amount to storage in cooling silos, and by the solidification of the waste through a vitrification process, its volume is reduced and rendered more impermeable to leaching by water. Dumping in the ocean of high level waste is now forbidden by the International London Convention of 1975, and so the only satisfactory solution is final storage in an extremely stable geological structure.

Various geological structures are being considered for their suitability, but only one site is currently underoing detailed experimental testing for this purpose. This is situated in the vast salt deposits at Asse in West Germany, where there has been no water penetration for millions of years. We conclude that until similarly satisfactory sites have been created in or outside the Community, there can be no claim to a proper solution having been found for the disposal of high (and, to some extent, intermediate) radioactive wastes. In this situation, there is a case for limiting the expansion of nuclear capacity. An installed capacity of 100/150 GWe would represent a reasonable maximum.

Nuclear power station operation

The chances of death caused by a release of radioactivity into the atmosphere from a core melt-down following a runaway chain reaction, have been estimated by independent bodies (UK Royal Commission on Environmental Pollution, 6th Report 1976 and the Ford Foundation MITRE Study 1977) at between 1 in 10^6 and 1 in 10^8 per reactor year. These figures are not based on actual experience, but on careful examination of the design of components of nuclear power stations. With operating experience, some further reduction in risk is likely, although it may be costly to achieve any significant improvements.

Applying these risk probabilities to the alternative energy strategies the risk of such an accident occurring by the year 2000 in the minimum nuclear strategy is estimated at approximately 1 in 600 - 1 in 60,000, and this reduces to 1 in 350 - 1 in 35,000 in the maximum nuclear programme strategy. (For comparison, the probability of death for an individual per single year of exposure is estimated as 1 in 400 from smoking ten medium tar cigarettes** a day and 1 in 8,000 from a traffic accident (UK)). The risks to the public from such a nuclear accident is less than that which is likely to be caused by the activities of many other industries and far less than those risks to the workforce, compared with other energy producing industries. The risk numbers given do serve to demonstrate how the risk of an accident (assuming pessimistically no improvement in safety standards) of a

---

* Plutonium -239 has a half life of 24,400 years.

** The implications of the accident of the Three Mile Island nuclear power station are discussed in a footnote on page 90.

large commitment of changes over time, and the relatively little change in the risk probability between the maximum and minimum nuclear programmes by the year 2000. This difference would become much more significant as time proceeds. There is no doubt that public reaction to this kind of accident would be on a far greater scale, and so inherently damaging to the Community's energy plans, than from the more likely accidents associated with other energy industries.

Terrorist risks

In theory, it is possible to manufacture and detonate a nuclear device from enriched reactor grade fuel*. Given the lack of resources likely to be available to terrorists, it is much more likely that any attempt to acquire fissile material, either by undercover agents from a plant or by hijacking the material in transit, would be directed towards plutonium. The availability of plutonium is directly related to the number of reprocessing plants built.

Therefore, we conclude that:

i. the ordering of reprocessing plants should proceed very slowly, allowing for a five year test period to follow on the building of the first one or two plants so as to allow operating experience to improve security techniques, even assuming a successful test period;

ii. the overall number of reprocessing plants should be limited in number; it is difficult to specify an exact number, but there should be a continued assessment of the risks of processing (and the economic advantages) vis-a-vis those of storing and finally disposing of unprocessed irradiated fuel.

If the number of reprocessing plants were limited, the associated security operation, involving the careful vetting of personnel and the presence of an armed security force, would not need to infringe upon public civil liberties in a significant way. The potential size of the security force and the extent of their activities can be minimised by improvements in design of plant and transport facilities, and doctoring of fuel.

Proliferation

The risk of nuclear proliferation cannot be entirely discounted when expansion of nuclear power capacity takes place, and undoubtedly, there is further room for improvement in the security procedures supervised by EURATOM. However, we believe that chances of proliferation occurring from a well-supervised EEC nuclear power industry is low in the 1976-2000 period compared with other methods that could be available to a non-nuclear nation bent on the means to manufacture a

---

* The United States Energy Research and Development Agency reported that in 1977 such a device had been detonated in the USA.

nuclear device. Again, the minimising of the number of reprocessing plants will make a major contribution to reducing this risk.

Fast-Breeder Reactor

The FBR itself is not considered any more prone to a core melt-down than thermal PWR stations. The coolant likely to be used in an FBR, liquid sodium, has been found to have a considerable robustness under severe conditions. In the event of an accident, the potential damage from an FBR is several times greater, both because of the speed with which it is likely to take place and because of the high concentration of the very toxic and radioactive plutonium.

The main concern associated with the expansion of FBR capacity is its requirement for plutonium as fuel. Not only would the potential terrorist and proliferation risk increase from a large FBR programme because of the essential requirement for reprocessing plant capacity and a resultant greater quantity of plutonium being moved about the country, but the FBR fuel fabrication process is somewhat more hazardous when reprocessed plutonium has to be added. Nuclear waste from an FBR programme would also contain a higher proportion of high activity material.

We would therefore conclude that no FBR programme should be embarked upon in the EEC until

i. there is a reasonable prospect that they are likely to prove as economic as alternative available energy supply options, including thermal nuclear power stations;

ii. a sufficient test period has been undergone for the initial fullscale FBR plant and for a fuel reprocessing plant;

iii. a satisfactory solution has been found for the disposal of high level radioactive waste.

- WHAT ARE THE IMPACTS OF EXPANDED FOSSIL FUEL PRODUCTION AND CONSUMPTION?

Production

The impact of expanded coal production in the Community will be marked in terms of local disturbance and visual amenity. The expansion of open-cast mining in West Germany could lead to direct confrontation with local residents as, in some cases, the operations will lead to whole communities having to be moved. In the UK, it is the intention of the National Coal Board to open up 3-4 major new fields in the 1980s and, they hope, a similar number in the 1990s. Social and environmental disturbance will not be as great as those arising from open cast or even previous deep mined operations, but in removing or minimising the effects of spoil heaps, significant local solid waste disposal problems could occur.

The expansion of offshore oil production in Europe, in spite of improved blow-out safeguards, will inevitably lead to further large spills, with possible damage to the littoral and marine zones. The most potentially serious spills could arise from an underwater crude oil pipeline fracture.

Two general conclusions can be drawn from the experience of the Ekofisk Bravo blow-out:

i. current oil-spill fighting equipment and platform operational safety standards are quite inadequate; steps are being taken to improve this situation;

ii. the potential environmental damage should not be over-exaggerated - the Norwegian Institute of Marine Research detected little damage to plankton beneath the surface of the sea area affected, and no detectable quantities of oil reached the Norwegian shore; the timing of the blow-out with respect to certain fish breeding patterns is a factor in determining impact.

It was assumed in all the energy strategies that indigenous fossil fuel production would be maximised.

Atmospheric pollution

The potential increase in atmospheric pollutants under the alternative energy strategies is given in Table 9.

Table 9

FORECAST EEC EMISSION (i) OF ATMOSPHERIC POLLUTANTS

In 10^6 tonnes/year strategies	1976 Actual	Increment								
		1985	2000							
			1A	2A	3A	4A	1B	2B	3B	4B
SO_2	16	2-3	4	8	3	8	3	6	2	7
NO_x	5	1	1	2	1	1	1	2	1	2
Particulates	3	-	1	2	1	2	1	1	-	2

(i) Assumes no additional control measures adopted and decreases sulphur content with increased consumption levels*.

Of these potential pollutants, the most serious is sulphur dioxide. The forecasts would imply a potential increase in SO_2 emissions of between 13% to 50% by the year 2000, of which the share from coal consumption is 70%-90%.

---

* The sulphur levels of imported coal are mostly lower in content than average EEC produced coal.

However, the amount of SO_2, NO_x and particulate local and regional deposition is not the same as the emission levels shown above. In a recent study, OECD concluded that 20%-60% of SO_2 emitted from the EEC is exported. Imports of SO_2 only offset this figure by about 5%.

Nor will the effect of increased SO_2 emissions be spread equally throughout the Community. The impact will be influenced by existing sulphur levels and the extent to which the sources are distributed or locally concentrated. In addition, topographical and climatic factors can combine to cause particular regional and local effects in determining ground level SO_2 concentrations, sulphate formation and 'acid' rain precipitation. Dispersion of emissions by the use of tall stacks has significantly reduced local and regional impact in some Member States, but is a major factor contributing to the export of SO_2.

It has been suggested that global weather patterns could be affected by carbon dioxide released from fossil fuel burning. CO_2 emissions are forecast to increase by between 10% and 45% by 2000. However, no evidence has so far been produced to substantiate conclusively that global climatic changes could occur, and by 2000, CO_2 emission from the Community is likely to represent less than 15% of total world man-made arisings.

Solid waste generation

The impact of solid waste generated from increasing coal production and use is local and regional. Taking Member States as a whole, there are sufficient locations to dispose of the 60%-80% increase in solid waste generated in the high fossil fuel strategies by 2000. This represents an additional 55-75 million tonnes/year of solid material, largely ash. However, it is very expensive to transport solid waste large distances and so there are strong economic pressures to find local solutions. The minimum expected increase in solid waste from coal combustion is about 30%.

The movement and disposal of this waste will undoubtedly cause local disturbances and could impinge on the amenity of the area. Additional problems will be created if wet limestone flue gas desulphurisation is introduced, as the residue from this process is a sludge, and is some 2-3 times the amount of solid waste produced with no flue-gas treatment.

- CAN POTENTIAL INCREASED ATMOSPHERIC POLLUTIONS BE CONTROLLED?

From oil consumption

The potential for additional environmental impact from increased oil combustion is relatively low

i. because overall consumption is unlikely to be increased by more than 37% by 2000;

ii. the means to control the most serious impacts are available and of reasonable cost.

Gasoline consumption by the year 2000 is likely to rise by between 20% and 40% depending upon the level of economic growth and the conservation methods adopted. While the effects and importance of NO_x and aromatic exhaust emissions must be continually reviewed, the potentially most serious impact is from increased lead emissions. Programmes to reduce lead in gasoline are being implemented or are under consideration in most Member States.

The potential increase in sulphur dioxide emissions from increased gas oil and fuel oil consumption, assuming no change in the average sulphur content of the fuel, would be 5% in the minimum fossil fuel (maximum nuclear) strategies and 50% in the maximum fossil fuel case. Even in the maximum fossil fuel expansion, a reduction of the average sulphur content of oil fuels by 50% could probably be achieved largely through the use of lower sulphur content fuels, and there would be no need for direct residual desulphurisation processes. The maximum increased cost (in 1976 prices) is estimated at \$6-\$12/ tonne of fuel oil consumed - about 7-15% of the delivered fuel cost.

Coal combustion

As pointed out, SO_2 emission from this source could increase by as much as 80% in the high fossil fuel strategies. Currently no commercial technique exists for removing sulphur from the fuel before it is burnt,unless it is converted to gas first of all. However, there are some promising experiments on direct desulphurisation of solid fuels taking place in the United States, The potential choice therefore lies between

i. flue gas desulphurisation;

ii. fluidised bed combustion.

The former method has been in existence for some time and is capable of removing about 90% of sulphur dioxide emitted. At present little flue desulphurisation takes place in the EEC. Various techniques exist, the most wellknown being the limestone scrubbing process. Cost estimates for installation of this sulphur removal technique in coal fuel power stations range from about 0.3-1.3 cents/kWh, although the most accepted estimates are around 0.25-0.35 cents/kWh. This represents a 15-30% increase on current coal generating electricity costs. These costs were based on an 0.6 station load factor, and can be up to 30%-40% higher at lower load factors. Flue gas desulphurisation costs, as a proportion of total fuel costs, are also markedly higher in industrial installations where the economics of scale do not apply.

Fluidised combustion technology is still being developed, though it is expected to be commercially available for use in industrial installations in 5-10 years. Its readiness for power station application is unlikely before 1990. Its attraction is that as well as removing sulphur dioxide (without forming a sludge) it also increases combustion efficiency, although the improvement in large power stations boilers would be small.

Overall then, there would seem a reasonable prospect that by 1990, when coal combustion could begin to increase sharply, control measures will be available to ameliorate the most serious atmospheric pollution impact without raising the real cost of delivered energy at that time by more than about 10-20%.

- WHAT ARE THE POTENTIAL IMPACTS OF RENEWABLE SOURCES OF ENERGY?

The development of renewable energy sources, as far as is known presents no health hazard to the Community. Nevertheless certain of them will present significant local environmental impacts.

i. Wave power - The installation of wave machines out to sea could affect the marine life in the littoral zones and present a potential obstruction to shipping and fishing.

ii. Tidal power - Tidal barrage systems are very large and are likely to lead to significant changes to the flora and fauna upstream of the barrage. These changes will occur in the water itself, in the littoral zone and possibly in the surrounding land through lowering of the water table. In some instances the effects could be favourable, but only careful study of each scheme could determine the impact with any confidence.

iii. Wind generators - Large wind turbines in exposed locations will present a considerable disturbance to visual amenity unless they are built out to sea. Some noise and radio/TV transmission interference can also result.

iv. Solar power stations - Although unlikely to make a very large contribution, heliostatic and photovoltaic solar power stations would cover land areas of an order of magnitude greater than those required for equivalent capacity thermal power stations.

v. Combined heat and power - The utilisations of CHP through the building of district heating systems involves the laying of extensive lagged pipework to housing complexes. This operation can take up to 15 years to complete and will cause significant disturbance to the urban environment in the process.

- WHAT ARE THE ENVIRONMENTAL IMPACTS ASSOCIATED WITH INCREASED USE OF ELECTRICITY?

Two general points are worth making about electricity generation.

1. The electricity supply industry is characterised by operations on a very large scale which have a number of very marked impacts on the environment. We are not persuaded by the arguments that the financial risks involved with building a few very large power stations, associated in a

situation of demand uncertainty are not yet sufficient to off-set the economies of scale and greater thermal efficiency of those stations (cf. a greater number of small plants). However, the utilisation of waste heat in CHP systems may reduce the optimum size of these power stations.

2. An increased unit of demand for electricity would not have to result in an additional unit of generating and transmission capacity being added, if that demand occurred at a time when there was spare capacity in the system, i.e. an improved load factor was achieved. Overall system load factor depends upon the nature of the market supplied, the climate and the way in which electricity is priced, and so varies from country to country. However, it is very difficult to predict how system load factor will change and there are as many arguments for believing it will deteriorate as there are for expecting it to improve. Therefore, it is assumed that a given rise in the overall demand for electricity will result in an equivalent rise in the generating and transmission capacity. The bulk of this new capacity will be supplied in large generating sets.

Environmental impacts of increased electricity demand are summarised in Table 10 below.

Table 10

ENVIRONMENTAL IMPACT OF FORECAST EEC ELECTRICITY DEMAND

	Increment 2000							
Strategies	1A	2A	3A	4A	1B	2B	3B	4B
Impact								
Approx. Installed capacity GW	349	258	313	282	250	235	267	245
Approx. No. of Thermal Power Stations (1)	220	159	182	161	154	144	146	134
Land requirement (2) hectares x 10^3	17	12	13	13	11	12	11	12
Water requirement (3) 10^9 m^3/year	5	3	4	3	4	3	4	3
Thermal pollution 10^{12} MJ/year	1.8	1.3	1.5	1.5	1.5	1.3	1.1	1.3

Notes:
(1) Assumes average size of 2000 MW.
(2) Requirement alters according to type of fuel and cooling system.
(3) Assumes 4.8×10^{-3} m^3/kWh for thermal nuclear stations and 2.7×10^{-3} m^3kWh for fossil fuel power stations.

Site requirements

The number of new sites required for thermal power stations (it is assumed that replacement of obsolete power stations will take place on the original sites, and although those sites will often not be suitable for the larger new power stations, this factor will be offset by the fact that fewer sites will be required) ranges from 110-200. This represents an average of 6 to 9 new power station sites having to be found every year. While this should not present insuperable problems in the next 10 years, suitable sites in the 1990s will become increasingly difficult to find, especially in the high nuclear strategies.

Water requirements

These sites will have to be located where access to cooling water can be obtained. It was assumed that none of these thermal power stations will be directly cooled by fresh water and that from 30-50% of nuclear stations will be cooled by sea water. In 1973/74, thermal electricity generation accounted for about 35% of total Community fresh water abstractions. Overall fresh water requirements by thermal stations may not rise by more than 10-15% by use of recirculating systems in new thermal stations, and by closing down of old directly cooled stations. However, in some areas of the Community, the rise in water abstraction requirements for this purpose could place a considerable strain on surface water supply.

Visual amenity

The visual amenity impact of increased electricity consumption results from:

i. the number of new power stations that will be installed; most of these will possess 2-6 large cooling towers;

ii. the building of new high and low voltage transmission lines in rural areas in approximate proportion to the increase in generating capacity - it costs 16 times as much to bury high voltage transmission lines than to erect overhead pylons.

Thermal pollution

Most of the waste heat will be transmitted to the atmosphere through use of cooling towers. However, some additional cooling water will inevitably be discharged into inland waterways.

The scale of environmental damage results from increased electricity consumption can be seen to be very large. The minimum expected growth for electricity would imply about a 30% reduced impact by the year 2000 below the maximum nuclear strategy. The environmental benefit of this minimum electricity consumption strategy would be seen mostly after 1985.

3. RECOMMENDATIONS

In developing recommendations for suitable energy strategies which would minimise future environmental impact resulting from having to meet the EEC's energy requirements in the period to 2000, we have also taken account of the following factors:

i. Cost of energy source - Those strategies for developing supply sources which would appear to have no chance of proving an economic proposition in the period of study, have not been considered.

ii. Uncertainty of future energy demand, supply and costs - Energy planning is particularly susceptible to uncertainty because of the nature, scale and long development lead times of energy developments. Flexibility should therefore be an essential ingredient in guiding future strategies, having in mind also the need to pursue options which minimise financial risk.

iii. Strategic implications - The report showed that, depending upon the alternative energy strategies adopted, the share of the EEC's energy supplies that is forecast to be imported in the year 2000 would range between 42% and 55%. The reliance on imported energy, as well as relating to the question of security of supply, can also have a considerable impact on the economic wellbeing of the Community.

It should also be pointed out that the recommendations are concerned with minimising future environmental impact through the adoption of appropriate energy strategies, and not with ameliorating the local impact of particular kinds of developments.

Recommendation 1: Energy conservation

The achievement of the maximum feasible levels of conservation in the various energy consuming sectors will require a much more comprehensive and vigorous policy to be introduced by Member States. This will involve subsidies, fiscal incentives, legislative measures, planning guidelines and R and D support. We suggest that such a policy can be justified on the grounds of

a. cost vis-a-vis alternative investment in new energy supplies;

b. conservation investment is not subject to the financial risks caused by uncertainty of demand, or by the scale and long development lead times of many energy developments;

c. it reduces import requirements, and could have some lowering effect on the long term OPEC oil price,

as well as leading to reduced overall environmental impact.

Recommendation 2: Minimising of electricity demand

Electricity in certain uses is essential and unavoidable insofar as it cannot be substituted by other fuels. However, alternative energy sources should be used wherever substitution can be made, particularly in space, water and process heat appliances, so long as large cost increases to the consumers do not result. By so doing, the several marked environmental impacts asociated with the electricity supply industry are minimised. Also:

a. waste of primary energy sources, inherent in thermal electricity generation is minimised;

b. there is reduced financial risk in avoiding commitment to large-scale capital projects with long lead times.

Consideration should be given to achieving this objective through appropriate pricing policies and/or Member States issuing guidelines on the uses for electricity.

Recommendation 3: Nuclear development

Four major points are made with respect to the rate of expansion of nuclear power capacity.

1. The rate of development should be restricted until a satisfactory solution has been found for the safe longterm storage of high level radioactive wastes.

2. There should be no rapid expansion of nuclear fuel reprocessing plants until sufficient operating experience has been gained on the initial fullscale plants that are built in the Community so as to ensure that this activity can be carried out safely, and also that the operation of the complete nuclear fuel cycle for AGR/PWR reactors does not present undue risks (including those associated with potential terrorist activity and proliferation). Given the advantages to be gained from limiting the number of reprocessing plants, there may be a case for irradiated fuel reprocessing being carried out on a Community rather than a national basis.

3. Research and Development should be encouraged into alternative fuel cycles, other than that of reprocessing, vitrification and long-term high level waste disposal.

4. Any development of Fast Breeder Reactors should be constrained by a proper test period for the initial fullscale plant(s), but also should only take place if the conditions set out in (1) and (2) above are satisfied. Also, because of certain increased risks associated with the FBR fuel cycle over those arising from thermal power stations, an FBR programme should only proceed if economically justified vis-a-vis other available alternatives.

5. Even if conditions (1) and (2) are observed, the nuclear power development should proceed at a measured pace with continual assessment of the overall risks (these are directly related to the number of power and reprocessing plants built). A commitment to expanding nuclear power capacity at the maximum feasible rate would constitute an inflexible supply strategy. As such it would make no allowance for the possibility of a change in the future relative costs of alternative energy supplies, and would be particularly susceptible to the consequences of public reaction to a nuclear accident.

Recommendation 4: Coal imports

We believe that there is a reasonable prospect that the potential environmental consequences, chiefly atmospheric pollution, from increased coal consumption can in the future be satisfactorily controlled within reasonable cost. While recognising that local disturbance and loss of amenity will occur from increased indigenous coal production, we do not feel these impacts are sufficient to offset the economic advantages this would provide (the same argument applies to indigenous hydrocarbon production) and the increased flexibility of supply. However, it is also likely to be necessary to import coal in increasing quantities if the policy recommendations on the rate of nuclear power are to be achieved. An active policy of securing coal imports should be pursued, and emphasis be given to encouraging the development of a suitable infrastructure for supporting the distribution of imported coal in the community.

Recommendation 5: Renewables development

In certain sectors, the use of renewable sources of energy can make an important contribution to meeting energy demands, albeit that this will have little effect on overall environmental impact by 2000. By so doing, increased supply flexibility would be achieved and reliance on imports and overall environmental impact would be reduced.

More importantly, we would recommend:

1. increased use of subsidies to Member States to encourage the installation of solar panels, chiefly for the domestic/commercial sector;

2. that consideration be given to introducing planning and building regulations so that new houses are built with the correct orientation for maximum utilisation of solar energy, and with higher insulation standards;

3. continual reappraisal of the economic feasibility of building district heating schemes to utilise combined heat and power from thermal power stations; the planning of new housing development should consider the possible use of such heating schemes;

4. financial support for and encouragement to industry to secure their involvement in the development of renewable sources of energy; without wishing to direct the flow of funds for this purpose, we would point to the potentially large contribution that wave power could make to the Community's energy needs, although careful assessment of the impact of utilising this energy source will need to be made, as well as appraisal of cost factors.

It is ERL's overall view that, by pursuing such recommendations as listed, the EEC's future energy requirements to the year 2000 can be met without an unacceptable increase in environmental impact, and would at the same time give proper consideration to the other factors which have to be taken into account in framing a suitable energy policy for the Community.

PART A—ENERGY DEMAND/SUPPLY

4. DEMAND FOR ENERGY

4.1 Introduction

It is necessary at the outset to define precisely the terms in which the demand for energy is to be considered in order that the reader should understand the relationship between the apparent requirement for fuel (and feedstock) in different sectors of the economy and the final demand for energy resources in the Community.

i. Useful energy - that quantity of energy actually required for the consumer to perform a particular function at a given level of activity, regardless of the form in which that energy is supplied;

ii. Delivered energy (heat supplied) - that quantity of energy supplied to a given consumer to meet his useful energy requirements; the relationship between i. and ii. is dependent on the type of consumer and the form in which energy is supplied;

iii. Secondary energy sources - these are the sums of the quantities of energy delivered to the various consumers in the form in which they consume it, e.g. coal, coke, natural gas, synthetic gas, oil products and electricity; the difference between the sums ii. and iii. is the loss incurred during the transportation process;

iv. Primary energy - that quantity of a given source of primary energy necessary to meet the sum of the secondary energy requirements; the difference between iii. and iv. are the losses incurred during any conversion process which takes place, e.g. in electricity generation, synthetic natural gas production, etc.

The loss differences beteen ii. and iii. are relatively small, usually less than 2%, and will be ignored in this study. All too often in prognostications of future energy needs, the other differences and relationships are also ignored and requirements for primary energy are discussed without any reference to what the energy is required for and the form in which it is likely to be supplied. Figure 1 on page 9. shows a chart demonstrating how the useful energy needs of the EEC in 1975, 465 million tonnes coal equivalent (including non-energy uses of oil), translated eventually into a primary energy demand of 1300 million tonnes coal equivalent (or 909 million tonnes oil equivalent). The chart also serves to illustrate the 'theoretical' potential for energy savings. However, it will be seen that much of this potential is not realisable, not only because of cost or feasibility, but also as a consequence of the inherent thermodynamic process taking place.

The purpose of the demand section of the report will therefore be to estimate the requirement for useful energy on a disaggregated basis in different sectors of the economy according to certain political and socio-economic assumptions. These will then allow an examination in Section 3 of the ways in which this demand might be met according to a set of different Community energy objectives and strategies. However, the two processes are interrelated to the extent that the demand for primary energy is, as was mentioned above, partially determined by the secondary energy form in which it is supplied. It will, therefore, be necessary to incorporate certain of the assumptions embodied in the energy supply strategies, e.g. the availability of natural gas, or the rate at which solar heating panels might be installed in households, into the estimates of demand for delivered energy.

4.2 Economic assumptions

There is ample evidence to support the belief that there is a direct relationship between the rate of economic growth (as measured by the Gross Domestic Product), and the demand for primary energy when viewed over a period of at least five years. This relationship is often expressed as the energy coefficient, defined as the percentage rate of annual increase in energy demand divided by rate of increase in GDP. The average energy coefficient for the EEC countries during 1960-1974 was close to 1.0, although there was significant variation between countries. This relationship is the macro-economic consequence of many micro-economic/physical patterns of behaviour within different sectors of the economy. These in turn are influenced by many factors, including the price of energy, level of industrial output, applied technology, industrial structure, demography, government laws, etc. All of these can change which in turn can alter the relationship between energy and economic growth, and so the necessity to examine the future demand for energy on a disaggregated sectoral basis is underlined.

However, the fact remains that, even recognising the many factors which influence the parameters of the energy/GDP growth relationships, the assumption taken on rate of GDP growth in the ECC economy is fundamental in determining the rate of future growth of energy demand. For the purposes of this study, whose purpose is to determine the potential social and environmental impact of alternative energy strategies, we believe that the differences of impact according to the strategy adopted are most likely to be highlighted by examining the effects of the alternative strategies applied to a single economic growth case. In selecting base case Community economic growth assumptions, GDP growth rates have been assumed which generally conform to recent middle-ground assumptions for the area as a whole:

1976 - 1985 GDP growth - 4.2% per annum

1985 - 2000 GDP growth - 2.7% per annum

The figure for the 1976-1985 period conforms to that shown as the weighted average economic growth rate for the member states in arriving at the energy balances for their national energy programmes, set out in the Second Report on The Achievement of Community Energy Policy Objectives for 1985, COM (77) 395 Final - July 1977(1). The growth assumed for 1986-2000 represents a middle order figure abstracted as the average of several long term economic growth scenarios, including those of World Energy Conference (2), the WAES Study (3), Belgrave of B.P. (4).

However, because of the sensitivity of long term energy requirements to the level of economic growth assumed for the period and the obvious uncertainty which pertains to this fundamental parameter, it is important that the implications for the various energy strategies of a significantly slower (or faster) rate of economic growth should be explored. As well as demonstrating the relationship between environmental impact and economic growth, this analysis should reveal those energy strategies whose inherent investment decisions and policies are most flexible and apposite (adopting criteria of economic return and security of supply) in the face of uncertainty about the growth of energy requirements. Thus, a low economic growth scenario for energy supply/demand is developed adopting GDP growth rates for the Community of 2.7% per annum for 1976-1985, and 1.8% per annum for 1985-2000. Similarly, the implications of a high GDP growth rate is also explored, in which 3.5% per annum growth (instead of 2.7% per annum) rate will be taken for the 1985-2000 period.

The other most important economic parameter influencing the growth in demand for energy is its price. We take the view shared by several energy economists that this price is likely to be chiefly determined by the price of OPEC oil in the period, and as such is dependent upon the relative supply/demand position for this commodity. Based on our assessment of this situation, we propose to assume that the real price of energy will be the same in 1985 as in 1976 (which is not to say that there will not be relative movement in the intervening period), and that the real price of energy will rise by 75-100% over the 1985-2000 period. (i)

In deriving the balance of secondary energy sources necessary to meet estimated sectoral delivered energy requirements, it will be necessary to have certain overall supply constraints in mind. The actual total availability of indigenous and imported energy sources are discussed in Section 5.2 of the report, which is concerned with energy supply.

(i) This assumption in itself is likely to prevent the 1985-2000 economic growth for the Community being sustained at the same rate as that for 1975-1985.

4.3 Conservation

Savings of delivered energy (heat supplied) can be envisaged as occurring in two ways:

i. price induced conservation - essentially this is the response of consumers, with some alerting by governments, to higher energy prices;

ii. government induced conservation - this refers to energy savings over and above price induced conservation which can be achieved as a direct result of deliberate government policy through legislation, education, fiscal incentives, grants and development planning.

In this report, it is proposed that conservation type i. should be built into the base demand estimates for the different sectors, whereas the higher level of conservation achieved through ii. is one of the energy strategies open to the Member States of the Community.

Forecasts for sectoral demand for delivered energy will now be developed.

4.4 Transport energy demand

This sector can be broken down into three main categories:

4.4.1 Road transport

This represents much the largest category, motor car and commercial vehicles between them consuming in 1975 72.0 and 26.6 million tonnes respectively. This accounts for 22% of all oil consumption, a figure which will rise in the next ten years.

Demand for gasoline is a function of car population,average annual distance driven per vehicle and average consumption per car. Little change is expected to occur in the second of the two variables (longer distances driven for leisure will be offset by an increasing number of two car families), but some consideration needs to be given to the other two factors. Car population for the Community trebled in the 1960's and is now predicted as growing as shown in Table 4.4.1 below,which assumes that EEC transport policy continues to cater for the motor car:

Table 4.4.1

FORECAST TREND OF CAR POPULATION IN THE EEC

	1970		1975		1985		2000
Car population (millions)	54.8		70.3		101.5		126.8
Average annual growth		5.1%		3.7%		1.5%	
Car population per 1000	217		271		373		440

The figure of 440 cars per 1000 head of population is slightly higher then the equivalent current figure for the United States, a country whose geography, higher standard of living and very poor public transport system, compared to EEC countries, is bound to give rise to a much higher car ownership than that for Europe. The forecast figure of 440 might, if anything, be looked upon as high, yielding a statistic for the year 2000 of about 0.8 car per head of population of driving age.

Reduction in vehicle fuel consumption may be realised by improvements in average efficiency of fuel use (on average about 20%) through better combustion and by manufacture of lighter cars and design change. Potential savings combining both these factors have been estimated as high as 50%. In spite of some statements to the contrary, a detailed study (5) on the effect of reducing overall carbon monoxide, hydrocarbon and NOX exhaust emissions by about 50%, and some reduction in current lead levels - as envisaged in currently proposed legislation - would increase the cost of new car manufacture by only 2% and no change in fuel consumption would result. If very stringent emission regulations equivalent to those in California were to be adopted, then it has been estimated that fuel consumption might increase by 10%.

As a result of the 1973/74 oil price increases, it is estimated that some 7-10% savings have already been realised by increased efficiency. By 1985 a further 6% overall improvement is expected, and a further 15% reduction by the year 2000. Adopting a vigorous energy conservation policy toward transport through subsidisation and expansion of public transport and legislation, a further 7-10% reduction in gasoline consumption could be achieved by 2000.

On the basis of these considerations future gasoline consumption is forecast as follows:

Table 4.4.2			
FORECAST EEC GASOLINE CONSUMPTION IN 1975-2000			
Million tonnes	1975	1985	2000
1. With price induced savings	72.0	99	109
2. With maximum conservation	72.0	97	102
3. With maximum conservation plus 15% of the market taken up by electric vehicles	72.0	97	87

Forecasts of gasoline (and other transport fuel) consumption at the low and high economic growth assumptions are shown in Appendix 4. It will be noted that the lower level of disposable incomes has much less effect on the projected gasoline consumption for the year 2000 than for 1985. This is simply a function of the flattening off of the gasoline growth curve in the 1990's which is expected in the base (and high) economic growth case.

Consumption of diesel fuel in commercial vehicles is expected to grow at a similar rate to the increase in industrial output and it is unlikely that much further scope exists for increased productivity through increased average vehicle size. A further 13-15% improvement in combustion efficiency should be achieved by the year 2000.

Table 4.4.3			
FORECAST EEC DIESEL FUEL CONSUMPTION IN 1975-2000			
Million tonnes	1975	1985	2000
1. With price induced savings	26.6	38	50.0
2. Maximum transport conservation	26.6	37	45.0

4.4.2 Air transport

Although air travel suffered a serious setback following the 1973/74 oil price rises, it is expected that it will resume its strong growth path during 1975-85, taken to be 5.0% per annum, a somewhat greater increase than that of economic growth. Air travel in the 1985-2000 period is expected to expand somewhat faster than the increase in GDP, although the rise in the price of oil is forecast to improve fuel/passenger productivity by 15%, chiefly through improved payloads. It will be noted that ERL does not take the view that developments in tele-communications in the next twenty years or so are likely to make much impact on the demand for personal travel.

Table 4.4.4

FORECAST EEC JET KEROSENE CONSUMPTION

Million tonnes	1975	1985	2000
	11.8	18.5	24.0

4.4.3 Inland water and rail transport

Greatly increased use of road transport has been at the expense of rail and, to some extent, water transport. No significant change in the demand for fuel in this category is expected unless Community transport policy alters radically.

Table 4.4.5

FORECAST EEC GAS OIL CONSUMPTION FOR RAIL AND WATER TRANSPORT

Million tonnes	1975	1985	2000
1. Price induced savings	5.6	6.0	8.8
2. Maximum conservation	5.6	5.9	7.0

4.5 Industrial sector

If one converts delivered energy back to primary energy, industry consumed about 38.5 million tonnes o.e., or 42% of total primary energy production. About two thirds of the total energy input to industry is consumed as direct heat or process steam, including electrical drive, the rest being accounted for by space heating, light and air conditioning use. Adding on the energy content of petroleum feedstocks increases total industry share of primary energy use to about 47%, although it is proposed that demand for feedstocks (excluding metallurgical coke) should be dealt with separately from that of energy. Clearly, changes in the pattern of delivered energy consumption in this sector will have a pronounced effect upon total demand for primary energy in the Community.

Just over 45% of total delivered demand is consumed by three energy-intensive industries - iron and steel, chemical and aluminium. Consequently, in projecting future demand, it will be necessary to consider them separately from other industries.

4.5.1 Iron and steel

Over the last 15 years, considerable improvements have been made in the energy consumed per unit of output and it is unlikely that the rate of improvement will be sustained. In the 1975-85 period, steel output for the Community is expected to grow more or less in line with GDP. After 1985, the likelihood is that growth will be somewhat less than that of the economy as developing countries (particularly OPEC nations) expand their steel industries. This increase in output will be achieved with less than the equivalent increase in coking coal consumptions as electric scrap furnaces take a larger share of total steel output and open hearth furnaces are phased out. In the long term, increased energy prices are expected to induce a 12% saving in energy from the current position and, if a vigorous government policy towards conservation is adopted, this could be increased to 20%.

4.5.2 Chemicals

This industry will continue to expand faster than the economies of the member states - about 6% per annum until 1985, thereafter at 4% per annum. Because they are high energy consumers, this industry is relatively energy efficient. A further 16% savings in energy consumed per unit of output is thought likely by the year 2000, which could be boosted by a further 4% through fiscal incentives.

4.5.3 Aluminium

Output of aluminium in the Community will probably be similar to that of general industrial output up to 1985; thereafter, the increasing price of energy will result in an increasing proportion of world aluminium smelting capacity being located in developing countries where cheap electricity is available. Substitution of more energy efficient processes (the most important of which being the emergence of an aluminium recycling industry) should result in a maximum of a further 18% energy conservation being realised in aluminium production.

4.5.4 Other industries

Historically, delivered energy consumption in this sector has grown at a similar rate to that of industrial output (which in turn is close to that GDP growth for the Community). Ignoring conservation efforts, this situation is likely to continue in the period to 1985. However, by the year 2000, it is to be expected that a 75-100% rise in real terms of energy prices, will result in a structural change in the development of industry in the Community towards less energy intensive activities. In addition to this there is substantial scope for energy conservation itself, even recognising that a substitution of coal for gas and oil and increased anti-pollution measures will work to the opposite effect. The OECD Secretary General is of the view(7) that 15-20% energy savings per unit output could be achieved by 1985, a figure which has met with general agreement in several member states. The longer term energy savings potential is estimated to be about double this level, although it should be pointed out that most of the potential savings are in fossil fuel use rather than in electricity consumption. If conservation were dependent only on the response of industry to price increases, it is not expected that energy savings by the year 2000 would be more than about 15-17%. With a vigorous government campaign, backed by legislation, education and fiscal incentives, this level should be increased to 25%.

Based on the considerations above, the following consumption of delivered energy in the Community industrial sector is forecast.

Table 4.5.1					
FORECAST EEC INDUSTRIAL ENERGY CONSUMPTION					
Million tonnes o.e.	1975	1985		2000	
		(1)	(2)	(1)	(2)
Aluminium	7	9	9	12	11
Chemical	39	62	62	103	99
Iron and steel	77	101	98	144	131
Other	149	198	190	256	232
Total	272	392	379	515	473

(1) Price induced conservation only.
(2) Maximum conservation government strategy.

The pattern of supply of different types of secondary energy sources to meet this industrial requirement for delivered energy depends upon the availability and relative price of different fuel types, which in turn can be influenced by the energy strategies of governments.

The alternative strategies pursued in this study are described in Section 3.2 of the following chapter on energy supply. The base case strategy 1A, which can be considered the extrapolation of the national programmes as projected in COM (77) 395 Final, is forecast to give rise to the following pattern of supply of delivered energy to industry.

Table 4.5.2					
FORECAST SUPPLY OF DELIVERED ENERGY TO EEC INDUSTRIAL SECTOR - STRATEGY 1A/1B					
Million tonnes o.e.	1975	1985		2000	
		(1A)	(1B)	(1A)	(1B)
LPG	4	7	6	14	14
Distillate oils	24	41	34	57	49
Fuel oil	66	100	95	118	104
Natural gas	56	93	93	75	94
Coal/coke	82	90	90	126	112
SNG	2	1	1	16	10
Electricity	38	60	58	109	90
Total	272	392	379	515	473

(1A) 1976 Community energy programmes/higher nuclear strategy; price induced conservation only.

(1B) Same as 1A, but with strategy of maximum government conservation policy.

The estimated pattern of supply under other energy strategies is set out in Appendix 2 at the end of the report.

Lastly, it remains to project the expected demand for non-energy oil products and petroleum (chiefly naphtha, but with increasing proportions of gas oil) feedstocks for organic chemicals under the economic assumptions adopted:-

Table 4.5.3			
FORECAST DEMAND FOR PETROLEUM FEEDSTOCKS FOR CHEMICAL INDUSTRY			
Million tonnes	1975	1985	2000
Base case	25	54	95
Low GDP growth	25	47	69
High GDP growth	25	54	100

The high 1975-1985 growth partially reflects the abnormally depressed 1975 consumption (1974 consumption was 34 million tonnes).

Table 4.5.4

FORECAST DEMAND FOR OTHER NON-ENERGY OIL PRODUCTS

Million tonnes	1975	1985	2000
Base case	23	30	33
Low GDP growth	23	26	31
High GDP growth	23	30	34

In the high economic growth case, the relationship between industrial energy requirements and GDP growth will be similar to that discussed for the base case, but with a trend towards greater structural change, implying lower energy consumption per unit of overall industrial output, at higher levels of GDP growth.

4.6 Domestic sector

This sector includes household, commercial, government department and agricultural delivered energy consumption. The latter represents about 4% of the sector demand. For households and other buildings, it is helpful to consider delivered energy as being made up of two categories:-

i. space and water heating use;

ii. electricity for lighting and appliances.

The latter is relatively simple and will be dealt with first. Consumption of electricity for lighting/appliances accounts for about 65% of electricity consumption in the sector and has grown over 1½ times as fast as the rise in disposable incomes over the last ten years, reflecting the fast rise in the gadget/consumer durables market and the improvements made to domestic and commercial lighting. It is expected that there will be a gradual flattening of this growth curve, mostly after 1985, as certain durables' markets approach saturation and also because of the slowing down in the growth of housing. However, it would be mistaken to overstate the degree of demand flattening, as previous forecasts of market saturation levels have nearly always been premature, and there will be some newer consumer markets, such as telecommunications, where electricity growth may well accelerate. Some scope (10-15%) for increased energy efficiency arising from consumers' response to increased electricity prices is to be expected.

Table 4.6.1			
FORECAST DEMAND FOR DOMESTIC ELECTRICITY IN NON-HEATING USES			
10^9 kwh	1975	1985	2000
Base case	286	646	973
Low GDP growth	286	560	697
High GDP growth	286	646	1040

It is in the consumption of energy supplied for space and water heating that by far the largest scope for energy saving exists through improvement of insulation standards. A study (6) carried out in the U.K. identified a potential reduction of 45% in the useful heat currently lost through walls and lofts, if high standards of insulation were adopted throughout the housing market, after allowing for a 2°C rise in house temperatures. In Western Europe as a whole, the maximum potential energy savings through insulation have been estimated as:-

wall insulation	-	25%
loft insulation	-	15%
draught proofing	-	7½%
double glazing	-	5%
		52½%

Improved overall burner efficiency (natural gas increasing its share of the market) will also contribute to a higher percentage of supplied heat being utilised.

These estimates of energy conservation potential in the domestic sector are considerably higher than was first considered to be the case. However, a recent OECD report (7) also suggested a figure of 50% was achievable. German and Dutch studies (8) (9) concluded that as much as 25-35% saving in the domestic/commercial sector was "technically possible and economically feasible at current prices" by 1985. ERL takes the view that without additional government encouragement and education, the figure is likely to exceed 11% before 1985 and even by the year 2000, savings induced by consumer response to energy price increases would not be more than 22%. However, with a comprehensive government conservation programme, it is believed that these figures could be increased to 16% and 40% respectively.

Growth in the demand for useful heat in the domestic sector is expected to continue to increase at not much less than the GDP growth until 1985. Thereafter, useful heat demand in the domestic/commercial sector is expected to rise at only 1.8% per annum as the rate of new house building falls.

On the basis of these considerations, the supply patterns of secondary energy to the domestic/commercial heating markets are forecast according to the Community energy strategies described in Section 5.3 and are set out in Table 4.6.2 overleaf.

It will be seen from Table 4.6.2 that under the maximum domestic energy conservation programme, supplied heat to this sector in the year 2000 is actually less than that in the year 1985.

The effect of lower economic growth rates on domestic energy consumption is shown in Appendix 4. As in the case of gasoline consumption, it can be seen that the lower level of disposable incomes does not greatly reduce the expected level of delivered energy requirements to the space heating market in the year 2000, since higher levels of personal disposable income gained through faster economic growth will result in the saturation level of the market being reached more quickly. (Saturation levels themselves may vary slightly according to the level of disposable family incomes, as the long term birth rate is likely to be influenced to some degree by economic growth.)

An observation might be made at this point. From the comments above on the effect of variations in GDP growth on energy consumed in the transport and domestic sectors, we are obliged to remark on the potential for error in making longer term energy forecasts, for it can be seen that demand in two out of the three main consuming sectors (currently accounting for about 50% of total primary energy consumption) depends upon some assessment of market saturation levels. Planners have generally been even less successful at forecasting these factors than assessing future levels of economic growth. The tendency to date has been to under- rather than over-estimate saturation levels.

Table 4.6.2

FORECAST DEMAND FOR DELIVERED ENERGY TO DOMESTIC HEATING MARKET *

Million tonnes o.e.	1975	1985	2000			
		1,2,3,4A	1A	2A	3A	4A
Coal	21	18	25	30	23	26
Natural gas	52	105	130	130	130	130
Oil	122	141	97	97	50	35
Electricity	12	10	29	12	18	17
CH and P	-	1	10	10	34	34
SNG	4**	2**	5	17	5	17
Solar	-	-	30	30	50	50
Total	210	277	326	326	309	309
		1,2,3,4B	1B	2B	3B	4B
Coal		17	25	30	20	26
Natural gas		105	111	111	111	111
Oil		131	65	60	20	20
Electricity		8	5	5	11	5
CH and P		1	10	10	34	34
SNG		2**	5	5	5	5
Solar		-	30	30	50	50
Total		264	251	251	251	251

A cases - price induced conservation only.
B cases - maximum conservation programme.
* includes commercial and agricultural demand.
** oil based synthetic gas.

5. ENERGY SUPPLY

5.1 Introduction

The purpose of this section will be to develop a set of different supply balances of primary energy sources to meet delivered energy demand as projected in the previous section. These balances will be determined according to a number of alternative energy strategies that could be adopted by the Community. It will, therefore, be necessary to define these energy strategies. However, in any of the energy strategies adopted, certain constraints on the availability of both indigenous and imported sources of energy will apply. These are now discussed.

5.2 Supply constraints

5.2.1 Natural gas

It is assumed that because of its clean properties and easy handling, the consumption of natural gas will generally be maximised, first in the domestic sector, then in the industrial sector and finally in the generation of electricity. Total consumption will be constrained by the availability of supply provided by indigenous production, supplemented by pipeline and liquefied natural gas imports. Based on reserve estimates in the Community and worldwide, and taking into account the competitive position and attitude of the Member States to securing gas imports, the following supply of natural gas is forecast for the Community.

Table 5.2.1

FORECAST NATURAL GAS SUPPLY TO EEC

Million tonnes o.e.	1976	1985	2000
Production	144	150	105
Imports	12	79	110
Total	156	229	215

It is expected that available natural gas supply will peak around 1990 at about 250 million tonnes o.e., after which this indigenous production will slowly decline.

5.2.2 Synthetic natural gas

The production of SNG based on coal is receiving considerable encouragement, particularly in the USA, UK and Germany. The number of full scale 2-2.5 billion cubic metres/year SNG plants operating by the year 2000 will depend upon the development of plant economics versus alternatives, and the energy policies of the countries concerned. The alternative strategies will be set out in Section 5.3, but here the point must be made that the availability of SNG will be limited by the R and D lead times involved, as well as the 3-4 year plant construction/commissioning time.

5.2.3 Oil

By the year 2000, crude oil is unlikely to be available in the way the Community has experienced to date. Domestic production, principally from the North Sea, is likely to level off and decline in the 1990's. The OPEC oil production ceiling is likely to be in the range of 40-45 million barrels/day, and the share of this production which the Community can be expected to receive, is unlikely to exceed the 1976 level. This would imply a maximum import availability of 600-675 million tonnes per annum. The following EEC crude oil production is forecast.

Table 5.2.2

FORECAST EEC CRUDE OIL PRODUCTION

Million tonnes	1976	1985	2000
Production	22	140	110

5.2.4 Coal

As with oil, there is a flexible upper limit on coal availability, determined principally by the rate at which world coal trade develops. Within an upper limit of 575 million tonnes of coal available for the EEC, it is expected that by the year 2000, coal will be the balancing energy resource. Maximum import availability is unlikely to exceed 250-300 million tonnes/year of coal. Indigenous EEC coal production is forecast as follows:-

Table 5.2.3			
FORECAST EEC COAL PRODUCTION			
Million tonnes hard coal equivalent	1976	1985	2000
Deep-mined	236	207	237
Open-cast	48	56	61
Total	284	263	298

Only in the UK is deep-mined production expected to increase, although it is not expected that the National Coal Board target in the Plan for Coal for 2000 of 170 million tonnes coal production will be fully achieved.

5.2.5 Nuclear

The supply, as distinct from environmental and political, constraints to the development of nuclear power are likely to arise principally in the area of skilled manpower. A private report "Nuclear Manpower to 2000",undertaken by the Energy Group at Newcastle University for the UK Atomic Energy Authority, concluded that, for the UK, an ordering programme of 4 GW/year could be prevented by insufficient availability of qualified nuclear scientists and engineers, but a more likely bottleneck was a general lack of skilled workers for such a large power station construction programme (some coal fired power stations also likely to be ordered). Another crucial factor influencing the productive capacity of the qualified nuclear workforce is the number of reactor types that national nuclear industries offer. In the maximum nuclear expansion strategy it is assumed that Germany will retain the BWR as well as the PWR capability, that the UK will build both AGR's and PWR's, whereas France is assumed to opt for the PWR only*,apart from its long term commitment to Fast Breeder Reactors.

The maximum construction capability of the Community nuclear power industry in the nine Member States (there will also be construction for export) is assumed to be 12 GWe/year from 1978-1985, and 15 GWe/year from 1985-1995. Of the total, Germany and France will possess an annual combined capacity of 7-8 GWe, the UK 2-3 GW, Italy 2GW and the remainder 2 GWe. Such a programme would yield an installed capacity of 293 GWe in the year 2000. It is assumed that Member States would be able to acquire sufficient uranium to support such a programme, although the price of uranium could double or triple from its 1976 price at $40-50/kg.

---

* FRAMATOME, the French nuclear power construction company, now has the capacity to manufacture all the heavy components of seven 900 MWe or five 1300 MW units annually (11).

The cost of nuclear power is a combination of power station capital and operating costs, the cost of enriched uranium fuel and the costs of irradiated fuel reprocessing, storage and disposal. The capital cost of nuclear power stations is obviously related to the reactor system installed. The estimates of capital cost range of a PWR thermal nuclear station given by Siemens, the German power station contractor, and the Nuclear Power Company, lay in the range \$550-\$700/kW, not including interest charges during construction, in constant 1976 prices. Adding to this interest charges at 10% over the construction period, the capitalised cost (assuming a 10% discount factor over 30 years) of enriched fuel supply and station operating costs of \$370-\$420/kW * (based on uranium ore costs of \$55-\$88/kg) and the capitalised cost of fuel reprocessing of \$185/kW (equivalent to £260,000/tonne of uranium **), gives a total capital cost of nuclear power of \$1,105-\$1,305/kW. It is worth remarking that the apparent estimated commercial cost of fuel reprocessing and storage has risen by over 100% in two or three years. Nor does this cost include that of the final disposal of high and low level radioactive wastes. So the conclusion is forced that until these costs are known and some operating experience of full-scale oxide fuel reprocessing has been gained, there is some uncertainty attached to the real long term cost of providing energy based on thermal nuclear electricity generation. Assuming a 60% load factor, amortising the capitalised cost over 35 years gives an electricity cost of 2.7-3.2 cents/kWh.

The Fast Breeder Reactor is unlikely to make any substantial contribution to energy supplies by the year 2000. France is building a 1200 MW FBR test station (Super-Phenix) which should be commissioned by the mid-1980's, but is unlikely to embark on the construction of any additional FBR capacity until 1990. The U.K. is to hold a planning enquiry in 1978/79 on the advisability of proceeding with its first full-scale plant CFR-1, and Germany, in collaboration with Belgium and the Netherlands, will only complete construction of a prototype station (SNF-300) in 1982/83. In view of the fact that no experience has yet been gained in building or operating a full-scale LMFBR, any assessment of the economics of such a plant will be subject to considerable uncertainty. It is probable that the price of uranium would have to increase by 2-3 times before FBR's were economic compared to thermal PWR reactors.

---

* Advised by British Nuclear Fuels Ltd to be 60% of combined fuel cycle and other operating costs shown in Table 11 of PWR capital/operating costs, published in "The Choice of Thermal Reactor Systems" (12).

** Evidence given by British Nuclear Fuels at Windscale Enquiry, U.K., 1977.

The supply constraints pertaining to renewable energy sources are discussed in the strategy assuming maximum development of these resources.

5.3 Energy strategies

In selecting strategies for meeting future energy demand in the Community, it is not the intention to focus on what might be the most appropriate or even likely number of alternatives available to the Member States. Rather the purpose is to identify those strategies which can be seen as the feasible extremes of different approaches to meeting energy demand, so that the political, social and environmental consequences of these different approaches can be clearly understood. This will subsequently help the reader to draw conclusions from the study.

The effect of lower (and higher) economic growth than that assumed in developing the main alternative strategies on energy supply will be examined for the base case 1A below. Since case 1A represents the highest energy consumption strategy because of the inherent assumption of high electricity growth, the high and low GDP growth sensitivities to this case necessarily yield higher energy consumption forecasts than would arise if the sensitivity analysis was performed on the alternative strategies.

Strategy 1A - national energy programmes to 1985/high nuclear expansion to 2000

This strategy can be viewed as a base case. The estimated supply balance of the combined national programmes of Member States to 1985 is set out in The Second Report on The Achievement of Community Energy Policy Objectives for 1985, COM(77) 395 Final. While not specifically stating it in these terms, this programme can be seen as a high nuclear growth strategy, insofar as increases in nuclear power capacity were seen in the 1976 national policies of most Member States as a major means towards containing the rise in oil consumption and limiting the increase in imports of crude oil - the two major energy objectives of the Community. This case is, therefore, the basis of the high nuclear growth strategy.

It is assumed that in pursuing this high nuclear programme, the necessary fuel enrichment and fuel fabrication capacity would also be expanded to meet the necessary demands of such a programme. In the long term, it is likely that a high proportion of the Community's uranium fuel requirements will be enriched and fabricated within the Community. The question of nuclear fuel reprocessing is rather more open to question and will be discussed in the section dealing with impact of different nuclear strategies. However, in this high expansion case, significant expansion of the Community's fuel reprocessing capacity will be unavoidable, as will the need to develop nuclear waste disposal sites. Most capacity is likely to be of the Pressurised Water Reactors and Boiling Water Reactor type, though some additional Advanced and High Temperature Gas Reactors are likely to be installed.

Strategy 2A - maximum fossil fuels

This option assumes that imports of crude oil and coal into the Community will be maximised within the limits of availability referred to in section 5.2. This strategy can be looked upon as the conventional alternative to that of maximum nuclear expansion. It is also the strategy which assumes that expansion of coal-based Synthetic Natural Gas plants will be maximised. This technolgoy is currently being developed with some urgency, particularly in the U.K. (where a small pilot plant is operating) and Germany.

It is expected that a full-scale (about 1.4 million tonne o.e./year output) plant will be in operation by the mid-1980's. The capital costs of such plant is such as to require currently an estimated ex-works gas price in the range of $4-6/million BTU's, over twice the current bulk price of heating oil but of the same order of magnitude of estimates delivered cost of Liquefied Natural Gas from new methane liquefaction plant in the Middle East.

Strategy 3A - maximum 'renewables'/maximum nuclear power

By 'renewables' it is meant those primary sources of energy which have no resource constraint, and which are thought to be generally environmentally attractive. They include solar energy, wind, wave, tidal and geothermal power. We also add to this list the utilisation of waste heat by means of Combined Heat and Power plants (CHP) and heat pumps which draw on the specific heat of some form of energy sink (such as drains, rivers or even the ground) by converting the low grade heat to high grade heat through the expenditure of mechanical energy. It should be said that all of these 'renewable' energy sources are now receiving development finance to a lesser or greater extent in the Community. A tremendous amount of work is required in some cases to bring their technology to a stage where they might be commercially viable when compared with alternatives. Some estimated costs will be mentioned in their brief descriptions below, but it should be emphasized that the potential for change in these costs compared with other energy alternatives over the next twenty-five years is enormous. For example, some dramatic reductions (2-3 times) in the costs of producing solar voltaic cells have been made in the last three years since increased research effort has been directed towards this end. Equally, as development proceeds, it may be found that the real costs of certain renewable energy sources are significantly higher than those indicated overleaf. The availability of 'renewables' in this energy strategy by the year 2000 is, therefore, assumed to be constrained only by the feasible rate at which development is considered to be able to take place, not by any assessment of economics.

(1) Solar

There are two basic methods for harnessing solar energy for provision of heat. The first is through the installation of absorption panels in the roofs of buildings, through which a heat transfer fluid is passed to distribute heat, either for space or water heating. Because of the lack of regular sunlight in northern Europe, solar panels are only capable of supplying around 10% of a building's space heating requirements (as long as some storage were available), whereas it could be expected to supply around 50-70% of water heating needs in a building of the correct orientation - this is an average for the whole of Europe. About 25% of the domestic sector's useful heat requirements is for water heating, 55% for space heating and the rest for lighting, appliances, etc. Also to be taken into account is the fact that only 40% of buildings today have the correct orientation for solar panels to be effective. By a deliberate planning, this figure could be increased to about 60% by the year 2000. Assuming that half of all buildings with the correct orientation for solar panels have them installed by the year 2000, then the contribution of solar energy to domestic heating would be:-

for space heating - 2%

for water heating - 8%

Their share could be supplemented by the contribution of 'passive' solar heating achieved through careful design orientation, window insulation and regard to the thermal mass of new buildings. It is estimated that 30-40% of a building's space heating could be provided in this way. ERL consider that new commercial buildings offer the greatest potential for 'passive' solar heating of this kind. Assuming that one third of new buildings were oriented and designed correctly from this point of view, then another 4-5% of domestic heating requirements could be met by solar energy. Overall then, it is estimated that solar energy could provide about 14-15% of the housing market by the year 2000. Where a high level of housing insulation was adopted, this contribution could be nearer to 20%.

While only marginally attractive from an economic standpoint at today's energy prices, the rise in real terms of energy prices expected after 1985, together with some fiscal incentives, could make domestic solar heating a relatively good financial proposition in the last fifteen years of the century.

A recent study carried out for ERDA in the United States calculated that solar panel systems are currently capable of providing industrial process heat up to 180°C (13). This would cover 35% of industrial process heating applications, a figure which would be nearer to 60% by the year 2000. The scope in European industry is rather less, but, with sufficient incentive, some contribution from solar energy to industrial fuel needs * could be made by the year 2000.

---

* Steam process heat accounts for about 35-40% of total industry energy consumption.

In spite of great improvements in conversion efficiencies of photovoltaic silicon cells in the last three years, we do not consider that electricity generation from this technology will make a significant contribution before the year 2000. The chief drawback is insufficiency of direct sunlight in EEC countries and the need for large power storage. Similar problems afflict the large scale development of the utilisation of solar energy to generate electricity by focusing sunlight using a heliostat mirror system. The combined contribution of these technologies in France and Italy (the countries with most potential in this respect) is considered to be not more than 1 million tonnes o.e.

(2) Wind power

Windmill electricity generation is considered to have some potential in certain of the EEC countries, where they could be located out to sea, chiefly France, Scotland, Ireland and, to a lesser extent, Germany, The Netherlands and Denmark. A joint report (13) by the Danish Academy of Technical Sciences and the Ministry of Commerce claims that at least 10% of Denmark's energy needs in the 1990's can be met by wind power. Considerable technical progress is still required to improve efficiency and reliability. It is estimated that at most 1000-1500 windmills of 2-5 MW generating capacity could, in theory, be established by the year 2000. Grid management problems are unlikely to arise with such a low overall proportion of wind power in the electrical generation/ transmission system. Estimates of the capital cost vary from $400/kW* installed for a high performance machine on a good site (14) to $700/kW (15). Estimates from the U.S. are towards the lower end of the range.

(3) Wave power

While wave power is considered to have considerable long term potential than that of the tides, the technology for exploiting it is much less advanced. The height of the wave is used to drive a hydraulic motor (the two most well-known designs currently being the Salter duck and the Cockroft raft, although there are four other systems under examination), which in turn drives a shaft geared to an electrical generator. It is hoped to have a prototype wave machine in operation by 1983 and, following successful demonstration, some 5000 MW of capacity could be installed in the U.K. by 2000. Similarly, France could also have a certain amount of wave power capacity installed by that year, and Ireland could perhaps contribute 1000 MW. To make full use of this capacity and to improve the economics of such projects, it would be necessary to construct additional pumped storage capacity, or manufacture and store hydrogen on site. The total maximum potential contribution

---

* Adjusted to 1976 prices.

of wave power in the EEC by 2000 is estimated at 10 million tonnes o.e./year. A considerable increase in the real price of energy or a lower interest rate in discounting future earnings on the capital employed on the projects would be necessary to justify the development of wave and tidal power. Average power density, with 25% overall conversion efficiency, for Salter ducks is estimated at 90 MW/kilometre (16). Current cost estimates in 1976 prices have been put at $800-$1500/kW. However, in view of the large offshore structures that would be necessary to spoort wave power machines and the high cost of underwater cables, it must be recognised that wave power development would carry high transmission and maintenance costs compared to nuclear power and other renewable energy sources. In calculating generating costs relating to the capital cost range given for average power densities, a 45% load factor was assumed.

(4) Tidal power

The Severn estuary provides the best potential site in Europe, where the erection of a long barrage could provide anything between 4.5 to 10 GWe, which could yield a contribution of 4 million tonnes o.e. Taking into account the other potential sites in the Community - principally at Mont St. Michel, France (Chaussey and Minquiers projects)- there would appear to be a maximum of another 25 million tonnes o.e. per annum available tidal power capacity for exploitation, of which 7-13 million tonnes o.e. is assumed to be completed by 2000. Construction time is likely to be about 15-20 years for a large tidal barrage and installation of turbines.

The U.K. Central Electricity Generating Board carried out in 1975 a study of the economic potential for the Severn Tidal Barrage scheme (17). On the basis of capital costs, expressed in 1976 prices of about $700-$750/kW, and a 10% interest rate (a crucial factor in a project of such a long construction time), the CEGB concluded that the project was economically unattractive vis-a-vis nuclear alternatives. However, in 1977 the Select Committee recommended (18) that a thorough reappraisal be undertaken. EDF in France have so far also argued that the Brittany tidal power schemes are uneconomic versus nuclear power. Load factors are likely to be in the region of 23-27%.

(5) Geothermal

Exploitation of hot rocks by pumping water through drilled holes 2-10 km into the earth's crust, and then utilising the resultant steam to generate electricity, probably has only a limited potential in EEC countries. At most, it is expected that France and Italy could expect the equivalent of an additional 1-2 million tonnes o.e. of thermal generating capacity. Updating previous estimates of geothermal power costs (19) and (20) would imply 1976 capital costs of $600-$1400/kW.

(6) Biomass and waste materials

The economics of the collection of biomass, whether in the form of timber, forestry waste or leaves, for subsequent conversion to hydrocarbon fuels through alternative bio-synthetic processes, is so unattractive as to virtually rule out any significant contribution of biomass to the EEC's primary energy sources by 2000. The combustion of waste-derived fuel in commercial boilers is already practised. A doubling of the real price of energy could provide a maximum EEC potential from waste of 10-12 million tonnes o.e., either directly burnt in incinerators or combusted as manufactured fuel.

(7) Combined heat and power

The utilisation of the 65-70% waste heat emitted from thermal power stations in district heating schemes could radically improve the thermal efficiency of conversion of electricity generation. In high density housing areas in Denmark and Germany, there are already successful schemes in existence with plans and financial assistance for further expansion. A recent study in the U.K. (21) identified a potential district heating market of 20 million tonnes o.e. in this country. In the EEC as a whole, it is thought that a potential of 100-125 million tonnes o.e. exists. However, in view of the long lead times necessary (about 15 years), it is probable that not much more than 10% of the domestic/commercial heating market could be supplied by 2000, i.e. about 30-35 million tonnes o.e.

(8) Heat pumps

By the use of electrically driven heat pumps, the conversion efficiency of primary energy is increased by 2.1 times over that of supplying electricity for direct use in space heating through the utilisation of low grade heat in the surrounding environment. The efficiency of conversion is related to the ambient temperature, and so heat pump installation is most suited to milder climates. Germany is already promoting the use of heat pumps through subsidies. Assuming that 10% of the domestic/commercial space heating market is supplied from heat pumps by the year 2000, then the effect of this development is to reduce by 5% the total amount of secondary energy sources that need to be delivered to this market.

It should be mentioned that, if large increments (over 10% of total generating capacity) of wind, tidal or wave power generated electricity are to be absorbed into the national generating grids, energy storage systems of sizeable capacity will have to be built to maximise capacity utilisation. This storage could take the form of pumped storage schemes, or else hydrogen or ammonia manufacture.

Strategy 4A - maximum 'renewables'/high fossil fuels

In this alternative, maximum development of renewable sources of energy will be supplemented by a high fossil fuel supply assumption, and, therefore, a lower nuclear growth.

Strategy 1B - maximum conservation/high nuclear power

The energy strategy in this alternative is identical to 1A, except that it will be coupled with government policies of maximum feasible conservation as identified in the sectoral delivered energy demand projection in Section 4.

Strategy 2B - maximum conservation/high fossil fuels

Again, this alternative mirrors 2A, except that maximum conservation policies are also assumed.

Strategy 3B - maximum conservation/maximum renewables/high nuclear power

This strategy can be looked upon as one which would minimise the use of fossil fuels in the EEC.

Strategy 4B - maximum conservation/maximum renewables/high fossil fuels

This strategy can be considered as the minimum nuclear expansion strategy.

It should be pointed out that when moving from one strategy to another, it is not always possible to substitute entirely one fuel type by the full availability of another, whose supply is being maximised, because of constraints within the market. For example, it is not possible to cut electricity consumption below a certain level in industry - see Section 4.5.

5.4 Energy industry use

This section will estimate the quantity of primary energy required in meeting the demand for secondary fuels identified in Section 2.

5.4.1 Electricity generation

The forecast consumption of electricity in transport, domestic and industrial sectors is as follows:-

Table 5.4.1

FORECAST OF CONSUMER ELECTRICITY DEMAND IN THE EEC

10^9 kWh	1975	1985	2000
Strategies 1A	982	1640 (5.4%)	2820 (3.7%)
2A	982	1640 (5.4%)	2340 (2.5%)
3A	982	1640 (5.4%)	2630 (3.3%)
4A	982	1640 (5.4%)	2465 (3.0%)
1B	982	1580 (5.0%)	2299 (2.3%)
2B	982	1580 (5.0%)	2220 (2.2%)
3B	982	1580 (5.0%)	2404 (2.7%)
4B	982	1580 (5.0%)	2273 (2.3%)
Low GDP growth case	982	1429 (3.8%)	1821 (1.7%)
High GDP growth case	982	1640 (5.4%)	2930 (4.0%)

Note: Figures in ( ) indicate average annual growth rates from previous year shown.

The minimum associated electricity consumption growth associated with the base case economic assumptions made are, from 1975-85, 4.8-5.2% per annum, and, from 1985, 2.2-2.4% per annum. The slow-down reflects both the lower level of GDP growth in the second period, and also the increasing saturation of the domestic appliance market.

It is also of interest to note that the 'renewables' strategies 4A, 3B and 4B, require a higher electricity growth than their corresponding non-renewable strategies, 2A, 1B and 2B, in order that their potential can be realised.

In the period 1975-85, there will be little overall improvement in average conversion efficiency in electricity generations as improved generating efficiency of thermal stations will be offset by the smaller relative contribution of hydro-power. A small 5% improvement of efficiency is assumed over the 1985-2000 period as the latter effect described is not sufficient to offset the overall improvement of thermal efficiency. Use of waste heat is shown in the consumption of secondary energy sources in the domestic sector - see Table 4.6.2.

The supply balance of primary energy to generate sufficient electricity to meet the above demand is shown in Appendix 2.

5.4.2 SNG manufacture

The forecast primary energy input to meet SNG demand as described in Section 4.2 is shown as follows:-

Table 5.4.2

FORECAST COAL DEMAND IN EEC SNG MANUFACTURE

million tonnes o.e.	1975	1985	2000
Strategies			
1A/3A	-	-	31
2A/4A	-	-	63
1B/3B	-	-	23
2B/4B	-	-	38
Low GDP growth (1A)	-	-	31
High GDP growth (1A)	-	-	55

These forecasts assume a 65% efficiency of conversion.

5.4.3 Refinery fuel requirements

It is initially assumed that total oil product demand will be met from refining capacity within the Community. In practice, of course, there will be imports and exports. It will be of interest to note that only in the high fossil fuel cases, 2A and 4A, will additions to the Community's greenfield site refining capacity be required. The consumption of refinery fuel is projected according to the various energy strategies in Appendix 3. Some allowance has been made for the increased percentage use of refinery fuel per tonne crude oil throughput caused by a greater proportion of catalytic reforming and cracking plant in total refining capacity.

Lastly, it should be mentioned, although there is a loss of primary energy (coal) resulting from the manufacture of coke, the coal/coke consumption shown in industrial energy demand already allowed for this conversion loss.

5.5 Primary energy balances

By summing the primary energy consumption by type derived from the projections of sectoral demands in the previous section, and adding the associated quantity of primary energy necessary to provide electricity, SNG and refinery fuel identified in Section 5.4, the total primary energy that must be supplied is arrived at according to the alternative energy strategies. These are shown in Table 5.5.1 and 5.5.2.

Table 5.5.1

FORECAST 1976-85 EEC PRIMARY ENERGY CONSUMPTION

Million tonnes o.e.	1976	1985		
		1A/3A	2A/4A	1B/2B/3B/4B
Solid fuels	220*	236*	242	235
Oil	542	658	668	631
Natural gas	156	229	229	229
Hydro	25	34	34	34
Nuclear	21	140	124	124
Renewables	1	1	1	1
Total	969	1298	1298	1254
Average per annum growth		3.2%	3.2%	3.0%
1976-85 energy coefficient		0.76	0.76	0.71

* These figures are somewhat higher than those shown in the coal consumption figures for the national programmes in COM(77) 395 Final but conform to latest Eurostat coal bulletin for 1976.

Table 5.5.1 contd.

FORECAST EEC PRIMARY ENERGY BALANCE FOR 2000

Million tonnes o.e.	1A	2A	3A	4A	1B	2B	3B	4B
Solid fuels	320	400	325	381	280	360	275	370
Oil	715	742	641	740	630	680	557	637
Natural gas	215	215	215	215	215	215	215	215
Hydro	35	35	35	35	35	35	35	35
Nuclear	400	240	346	185	316	177	309	124
Renewables *	53	53	114	114	53	53	114	114
Total	1736	1685	1676	1670	1529	1520	1505	1493
Average per annum growth 1985/2000	1.9%	1.8%	1.7%	1.7%	1.3%	1.3%	1.2%	1.2%
Energy coefficient 1985/2000	0.70	0.66	0.63	0.63	0.48	0.48	0.44	0.44

* Does not include contribution of waste heat CHP and utilisation of low grade specific heat of environment by heat pumps. Their contribution is reflected in savings of other energy supplies shown.

Table 5.5.2

FORECAST EEC PRIMARY ENERGY BALANCE - LOW AND HIGH ECONOMIC GROWTH CASE

Million tonnes o.e.	1985		2000	
	Low	High	Low	High
Solid fuels	233	As 1A (above)	315	387
Oil	561		628	730
Natural gas	229		215	215
Hydro	35		35	35
Nuclear	97		163	400
Renewables	1		48	55
Total	1153		1404	1822
Average per annum growth	2.0%		1.3%	2.2%
Energy coefficient	0.74		0.72	0.64

5.5.1 Comparison with other long term energy forecasts

The forecasts of total primary energy consumption for 2000 in Table 5.4.3 lie within the range of the only other energy forecast produced for the EEC by CEPCO (10) which was 1400-2280 million tonnes o.e. Other sets of forecasts of primary energy consumption for 2000 have been projected for Western Europe. If one assumes that the EEC will consume 77% of total Western Europe energy demand in the year 2000 (cf 80% in 1976), then the following represent the consumption forecasts of other studies.

Table 5.5.3

COMPARISON OF ERL AND OTHER ENERGY CONSUMPTION FORECASTS FOR 2000

	ENERGY CONSUMPTION		GDP GROWTH	
	10^6 tonnes o.e. - 2000	1976/2000 growth	1976/2000	Average Energy Coefficient
ERL - base	1493-1736	1.8-2.5%	3.2%	0.56-0.72
- low	1404	1.5%	2.1%	0.71
WAES (3)				
- high	1725-1782	2.4-2.6%	3.8%	0.63-0.68
- low	1422-1529	1.6-1.9%	2.4%	0.67-0.79
B.P.	1671-2102	2.3-3.2%	NA	
WEC (2)*				
- high	1607	2.2%	3.6%	0.61
- low	1538	1.9%	2.7%	0.79

* World Energy Conference

While not strictly comparable, the ERL forecasts do not differ very greatly from the WAES projections, though the WAES projections assumed a larger contribution from nuclear energy and somewhat smaller shares from other sources. The World Energy Conference forecasts also bear quite close resemblance to the total primary energy consumption forecast by ERL, but they generally assumed a larger availability of coal and a lower supply of coal. ERL forecasts also assume that a somewhat higher level of energy conservation can be achieved through government action for a given level of economic growth.

5.6 Imported primary energy requirements

Based on the forecast of indigenous production of fossil fuels in Section 1, the following imported primary energy needs result:-

Table 5.6.1

FORECAST EEC IMPORTED FOSSIL FUEL REQUIREMENTS

Million tonnes o.e.	1976	1985			
		1A/3A	2A/4A	1B/2B/3B/4B	Low GDP Case
Coal	25 (11)	52 (23)	58 (24)	51 (22)	49 (21)
Oil	520 (96)	518 (79)	528 (79)	491 (78)	394 (70)
Natural gas	12 (8)	79 (34)	79 (34)	79 (34)	79 (34)
Total	557 (57)	649 (50)	665 (51)	621 (50)	522 (45)

( ) indicates % of total primary energy consumption.

Table 5.6.1 contd.

FORECAST EEC IMPORTED FOSSIL FUEL REQUIREMENTS FOR 2000

Million tonnes o.e.	1A	2A	3A	4A	1B	2B	3B	4B	Low GDP case	High GDP case
Coal	107	192	117	179	72	147	67	161	87	174
Oil	603	630	519	620	520	568	447	527	531	618
Natural gas	110	110	110	110	110	110	110	110	110	110
Total	820	932	746	909	702	825	624	798	720	902
% of total primary consumption										
Coal	32	48	36	46	25	40	24	44	32	45
Oil	84	85	82	85	82	84	80	83	82	85
Natural gas	51	51	51	51	51	51	51	51	51	51
Total	47	55	45	55	46	54	42	54	50	50

5.7 Comments on projected energy supply balances

The following major points can be made with respect to the forecasts of Community energy requirements and to the alternative supply strategies developed to meet that demand.

i. Assuming 4.2% per annum GDP growth 1976-85, a 2.7% per annum growth 1985-2000, and a doubling of energy prices in real terms, the requirement for primary energy is expected to rise between 1.5 and 1.8 times. This is achieved through a gradual lowering of the energy co-efficient; 0.71-0.76 to 1985 and 0.44-0.66 from 1985-2000. This compares with the 1960-74 figure of 0.96.

ii. Under a low economic growth scenario (GDP increase 2.7% per annum 1976-85, 1.8% per annum 1985-2000), it is estimated that energy demand would rise by 1.35 to 1.5 times by 2000; in a high growth situation (GDP increase 4.2% per annum 1976-85, 3.5% per annum 1985-2000), estimated energy requirements by 2000 would be 1.6 to 1.9 times the 1976 level. It will be noted that more efficient use of energy i.e. a lower energy co-efficient is expected at higher output of the economy; conversely, a larger energy co-efficient is likely to result in the lower GDP growth scenario.

iii. By introducing vigorous and comprehensive conservation policies, it is estimated that total primary energy consumption in 2000 can be cut by about 10% from the level that would otherwise arise.

iv. Renewable sources of energy could supply 105 million tonnes oil equivalent or 6-8% of total primary energy demand by 2000, if developed at their maximum feasible rate.

v. Electricity consumption is expected to grow at a significantly faster rate than overall energy requirements. In the maximum nuclear strategy, it is forecast that electricity consumption in 2000 will be virtually 3 times that of the 1976 level. The minimum growth that is likely under the economic assumptions taken, i.e. that which would result from restricting electricity usage to those consumers where no substitute fuel exists, would still give rise to an increase of 2.2 times that of the 1976 level.

vi. In order to avoid nuclear power capacity being expanded significantly beyond levels projected for 1985 (i.e. by more than 15-25 GW beyond the current programme envisaged for the Community), it would be necessary to adopt vigorous conservation measures, maximise the development of renewable sources, hold oil consumption at projected 1985 levels and increase imports of coal from present levels of around 38-55 million tonnes/year up to 200-225 million tonnes/year.

vii. Alternatively, in order to maintain combined oil and coal consumption at projected 1985 levels, i.e. around 900 million tonnes o.e., it would be necessary to expand nuclear power capacity by 2.5-3 times of the existing programme.

viii. Whatever energy strategy is adopted, it would appear that imported primary energy requirements will need to make up between 42% and 55% of total consumption compared with the 1976 figure of 57%. It is unlikely that the Community objective of 50% imported energy dependence can be achieved without expanding nuclear power capacity by at least 60-100 GW beyond that envisaged in the national programmes as set out in COM (77) 395 Final.

PART B—ENVIRONMENTAL IMPACT

6. INTRODUCTION TO PART B

In the second part of the report, we review the potential environmental impact of the alternative energy supply options. In Sections 7-11 we examine briefly the various impacts associated with each of the various primary and secondary sources of energy production and use, and examine to what extent these impacts may be ameliorated and at what cost. Clearly, the choice of site can also have considerable influence over the degree of local impact. However, as already noted in the introduction, such specific site factor effects are not the concern of this study.

As will be seen, environmental impact can be global, regional or local in nature, and it will be important to recognise these differences. The degree to which qualitative as well as quantitative assessment of these impacts can be made is usually found to be in reverse proportion to their spatial dispersement. However, the difficulty of determining the significance and extent of certain global-type impacts, on the lack of conclusive evidence on such environmental change, does not, of course, remove the cause for concern over such issues as potential climatic effects, etc.

In the final Section 12, we evaluate the impact in relation to the Alternative Energy Strategies developed in Part A of the report.

7. THERMAL POLLUTION FROM ALL SOURCES

The conversion and use of energy of any form results in heat losses which may be partially recovered and used. The waste heat may be discharged in cooling water (to inland waters or the sea) or into the atmosphere, either directly in the form of exhaust gases and warmed air from air-cooling systems, or indirectly by transfer of heat from the cooling water.

It has been suggested (22) that this continuing emission of increasingly large quantities of waste heat to the atmosphere will result in a long term rise in global atmospheric temperatures with consequential effects on the world climate. In 1975, the man-made contribution to the radiation balance was 9×10^9 kilowatts (23), equivalent to 0.02% of the natural radiation balance. Natural climate variables appear to dominate man-made perturbations. To date, there is no cast-iron evidence to support this hypothesis (24) but this may be partly due to the extreme difficulty of establishing a long term relationship of this sort.

Of more immediate concern is the possibility of the micro-climatological impact from large, geographically concentrated sources of thermal pollution and the effect of thermal discharge from power stations on the aquatic environment where 'once-through' cooling is employed. In the latter case, however, we anticipate that (with the possible exception of any new plants located on the coast) most new electricity generating stations will employ recirculating cooling systems. Here the problem of aquatic pollution is exchanged for the amenity impact of cooling towers and the possibility of micro-climatological effects. This will also be discussed in a later section dealing with the impact of thermal electricity generation. Also, in some areas of concentrated industrial/residential conurbations, the average heat output from man-made sources can dominate the natural flux and has been shown to cause more clouds, precipitation and higher temperatures than in similar rural geographic areas (23).

The quantity of waste heat produced by the alternative sources of energy production and consumption are compared below.

Table 7.1

WASTE HEAT GENERATED PER 1000 MW (HEAT) INPUT

In MW (Thermal)	Utilised	Waste [2]
Natural gas combustion (domestic/industrial)	800-900	100-200
Oil combustion (domestic/industrial)	700-850	150-300
Coal combustion (domestic/industrial)	650-770 [1]	230-350
High BTU coal gasification	550-680	320-450
Fossil fuel electricity generation	400-420	580-600
Nuclear electricity generation [3]	320	680

(1) Fluidised bed combustion would increase the high end of the range by some 5- 15%.

(2) Due to less than 100% combustion efficiency, these estimates slightly exaggerate the total heat emission.

(3) Based upon LWR system. AGR and HTR systems possess 41% and 39% thermal efficiency respectively.

It should, of course, be realised that both the waste and useful heat derived from fossil fuel combustion contributes to thermal pollution, whereas in thermal power stations, from 32-42% is converted to electricity. However, the point about power stations is that this heat emission is much more concentrated.

8. NUCLEAR POWER

8.1 Introduction

The utilisation of heat from nuclear power as a primary source of energy in the Community has, to date, been entirely in the form of electricity generation. Although some direct process heat application, in conjunction with electricity generation, is anticipated in future steel production, this will only present a very small proportion of total nuclear power output and the following description of the impact of nuclear power development is confined to its use as a primary energy source for generating electricity.

A later section will describe the impacts on the environment that are common to all forms of thermal electricity generation. Here we shall examine only those unique aspects of nuclear power which distinguish it from the production and uses of other primary energy sources.

There is a superficial similarity of nuclear power to the generation of power from fossil fuels, involving, as it does:

- extraction of raw materials;
- preparation of fuels;
- generation of energy as heat; and
- collection and disposal of residues from generation.

The presence of radioactivity in most process stages, but especially after the reactor, the major safety aspect of nuclear criticality and consequent explosion risk, and the presence of re-usable fuel material in generation residues with the probable need for recovery, all these combine to make the nuclear system markedly different from one based on coal, oil or gas.

A description of nuclear power can most usefully be given under three main headings:-

i. fuel cycle operation;

ii. nuclear reactor operation;

iii. social and political aspects.

8.2 The nuclear fuel cycle

The main stages of the nuclear fuel cycle are shown in Figure 8.2 This diagram assumes that all the processes necessary for maximum fuel utilisation are in operation, although, as will be discussed later in this section, to some extent the option exists for eliminating spent fuel reprocessing with its associated uranium and plutonium recycle stages.

The principal aspects of the fuel cycle stage are summarised in Table 8.2.1 overleaf.

Comments on the different stages of operation are given below. Before commenting on the different stages of the fuel cycle, the following general points are made:

1. Chemical hazards

Except for the manufacture of the highly reactive Uranium Hexafluoride, used in the enrichment process, involving the production of fluorine and hydrofluoric acid in the process, the chemical hazards of the fuel cycle are not very unusual. There are no other particularly toxic materials involved, other than those of the uranium and transuranium elements which will receive separate comment, nor are the processes difficult to operate. The reactions themselves are complicated by the presence of radioactivity, but not so as to affect the actual chemical hazards.

The effluents from the enrichment and fuel fabrication stages after 1GW year of opeation are:

- about 20 tonnes solids - heavy metal and calcium fluorides;
- about 25 tonnes liquids - off-gas scrubber liquors;
- 50 to 100 tonnes uranium metal as hexafluoride.

The depleted metal is potentially useful in a breeding programme.

2. Criticality

Criticality is a description of the circumstances under which the amount of fissile material is present in sufficient concentration, affected by size and shape, so as to allow a chain reaction to take place. In a nuclear reactor, the process is controlled so that the average amount of fissions, with its accompanying heat and radioactivity release, remains constant. In all stages of fuel enrichment, fabrication, reprocessing and storage, immense care has to be taken through the design of the plants and by excluding 'moderators' - in particular water - to prevent accidental criticality occuring. However, it should be said that the concentrations of radioactive U_{235} are never

Figure 8.2

THE NUCLEAR FUEL CYCLE

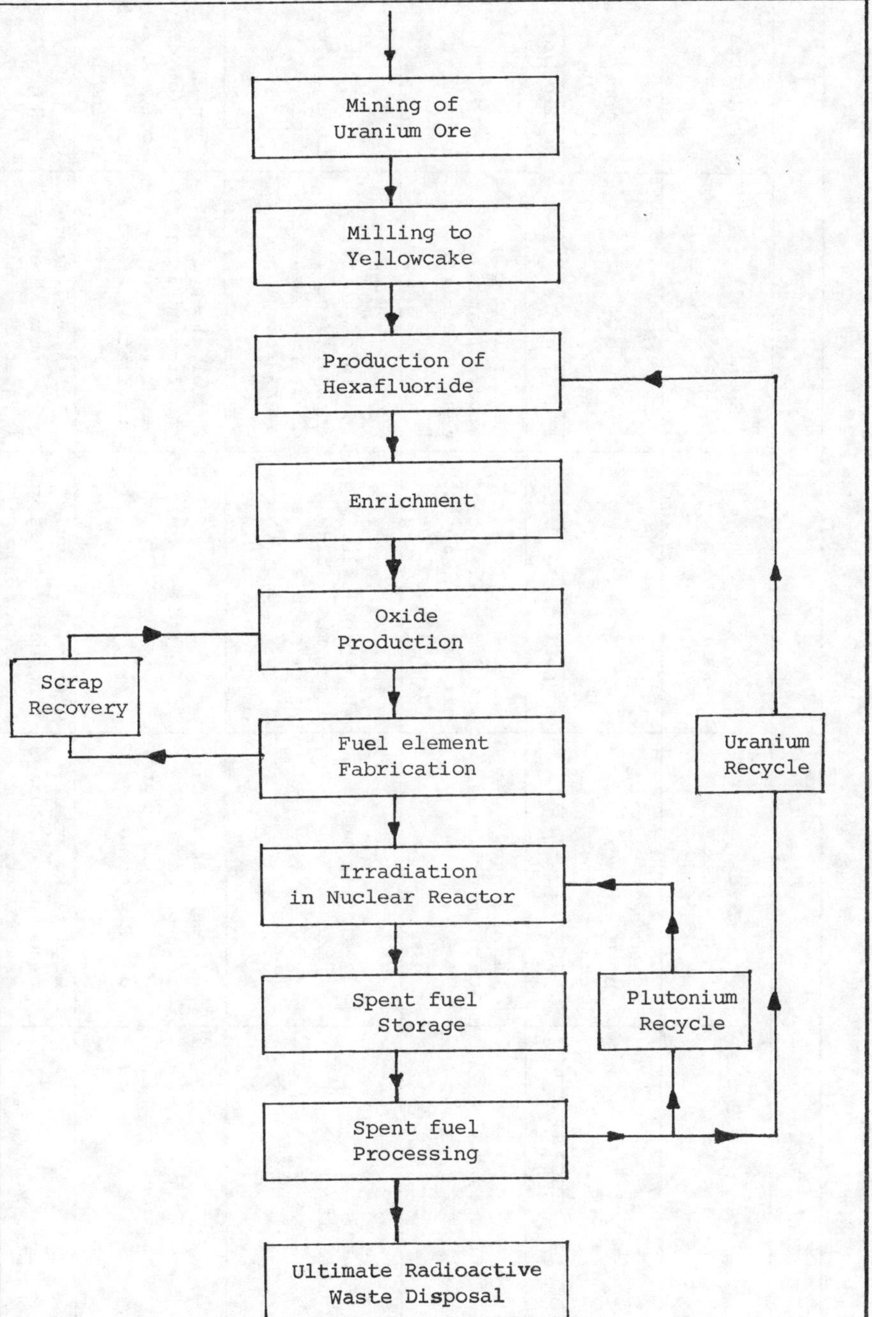

Table 8.2.1

FUEL CYCLE IMPACTS

Impact / Fuel Cycle Stage	Oxide Production	Fuel Fabrication	Spent Fuel Storage	Spent Fuel Transport
Chemical Hazard	Chemical activity of UF_6 before conversion.	None	None	None
Energy use	Negligible	Negligible	Negligible	Negligible
Effluents	Ammonia gas. Calcium Fluoride.	None if system is working well, as all material is too valuable to lose.	Resins from pond water cleaning. Waste heat from FP decay.	None
Radio-activity	No significant risk.	No significant risk when processing fresh material. Major problem when recycled U or Pu oxides are included.	Extremely high levels of activity in spent fuel, decaying somewhat over storage period.	Very high levels in unprocessed fuels.
Fire/ Explosion	No risk. Accidental release of UF_6 possible, but quickly condensed.	No risk	Radiolytic hydrogen from cooling water must be ventilated.	No risk. Problem of vehicle accident requires great container strength.
Criticality	High risk, requiring careful design and operation.	High risk, requiring careful design and operation.	No risk - elements only removed when incapable of sustaining fission.	No risk

Table 8.2.1 - Page 2

FUEL CYCLE IMPACTS

Impact \ Fuel Cycle Stage	Oxide Production	Fuel Fabrication	Spent Fuel Storage	Spent Fuel Transport
Safety and Security	Security measures essential for human safety, and to avoid criticality.	Fresh fuel requires some security - much more is needed if high Pu content is included.	Necessary only for personal safety. Fuel is far too dangerous to be hijacked.	Necessary only for personal safety.
Social Aspects	A small scale operation, unlikely to create a social burden. Transport of enriched U feed creates same problems as Pu.	A small scale operation, but with large security problems, creating a definite burden.	An essential part of reactor operation. Initial storage will be at reactor site - reducing attractiveness for in-town siting.	Fuel will have to travel on public roads or rail - so creating a definite burden, but security need is less than refined fuels.

Table 8.2.1 - Page 3

FUEL CYCLE IMPACTS

Impact \ Fuel Cycle Stage	Fuel Reprocessing	Recycled Fuel Transport	Wastes Transport	Wastes Disposal
Chemical Hazard	Wet processes use acids and organic solvents. Dry processes use fluorides.	None	None	None
Energy use	Negligible	Negligible	Negligible	Negligible
Effluents	FP gases. Low level wastes. Fuel element cladding. Scrubber liquids. High level wastes.	None	None	Decay heat.
Radio-activity	Very high levels at all stages of processing.	Majority of activity removed in wastes, but recycled fuels carry significant levels.	Very high levels.	Very high levels unaffected by subsequent processing.
Fire/ Explosion	Organic solvents and radiolytic hydrogen create definite risk.	No risk. Container must protect against vehicle accident.	No risk. Container must protect against vehicle accident.	No direct risk but decay heat could create fire in other materials.
Criticality	'Head end' has no risk, but as U and Pu are separated the risk returns.	Definite risk - transport and storage must be properly designed and operated.	No risk	No risk
Safety and Security	Essential on whole plant for safety of personnel, and on separated U Pu against theft.	U and especially Pu in transit are most vulnerable to 'diversion'. Security must be strong.	No risk	No risk
Social aspects	A fairly normal chemical operation vastly complicated by radioactivity and criticalit problems. Of value only if waste problem is solved.	Most serious impact will be need to improve massive security on cross-country runs, with increasing journey frequency.	No serious burden as volume is low.	'Safe' disposal not yet - and may never be - demonstrated. Public acceptance of active waste dumps inaccessible regions may be necessary. Use of wastes needed

Table 8.2.1 - Page 4

FUEL CYCLE IMPACTS

Impact \ Fuel Cycle Stage	Mining	Milling	Hexafluoride Production	Enrichment
Chemical Hazard	None	Use of sulphuric acid as leachant - not serious risk.	Significant use of F_2 and HF gases in high temperature processes.	High chemical reactivity of UF_6.
Energy use	Negligible	Small - about 0.1% of fission heat.	Negligible	The major step - consumes 4.4% of produced energy. Appears all as waste heat.
Effluents	Overburden from surface mines.	Negligible chemical effluents. Major discharge of solid tailings (99.6% of ore) and liquids to lagoons.	Liquids from dust and gas scrubbers. Some solid impurities as fluorides.	Only 'effluent' is major residue of depleted uranium. Waste heat.
Radioactivity	Evolution of natural Radon - no hazard in surface mining, but underground mines must be well ventilated.	Further release of Radon requires good ventilation of mill buildings. Tailings have natural uranium activity residue.	Negligible release in effluents. Risk increases when recycled U from reactors included.	No activity release. Depleted uranium has low activity level.

Table 8.2.1 - Page 5

FUEL CYCLE IMPACTS

Impact \ Fuel Cycle Stage	Mining	Milling	Hexafluoride Production	Enrichment
Fire/ Explosion	No risk	Use of organic solvent means definite risk of fire or vapour explosion.	No risk. Accidental release of UF_6 possible, but it condenses quickly.	No risk. Accidental release of UF_6 possible.
Criticality	No risk	No risk	No risk	Definite risk of critical assembly - needs care in storage design and water exclusion.
Safety and Security	No risk	Normal chemical plant safety requirements.	Safety requirements for hazardous chemical operation.	Production of enriched fuel requires strict security measures, as well as plant safety.
Social Aspects	Most uranium mines are far from point of use and in lightly populated areas.	Milling usually occurs close to mine, so major tailings problem does not become serious in developed areas.	A conventional chemical operation with no serious social impact.	Diffusion plant is large, noisy and unattractive. But centrifuge-based plant much better. Security measures could impose local burden. Waste heat problem.

sufficient at any stage of the fuel cycle or reactor operation to create conditions necessary for a nuclear explosion as would be achieved by a nuclear bomb. The main danger, if criticality were established, would be to the personnel of the plants themselves from the resultant gamma radiation and neutron emission. In the worst conceivable accident, wherein fracture of the containment system also occurred, a quantity of radioactive material might be released into the atmosphere.

8.2.1 Uranium * ore mining and milling

Only a relatively small proportion of the Community's uranium requirements is produced indigenously (in France) - just over 25% - and it is probable that this proportion will fall rather than increase. Uranium is, therefore, imported either in the form of 'yellow cake' (U_3O_8) which is the concentrated oxide produced by milling the ore (removes volume of tailings by about 90-98%) or, in the form of enriched fuel. It is probable that in the longer term the Community will import the majority of its uranium requirements as 'yellow cake'. This material is naturally radioactive but only slightly so, so that the principal concern lies in ensuring this importation is carried out under secure supervision to ensure that no 'yellow cake' is lost. Improved accounting procedures and supervision of this operation have been introduced by EURATOM since an incident in 1967, when 200 tonnes of U_3O_8 'disappeared' while en route for France - see later Section dealing with nuclear proliferation.

It is estimated that the operation of a 1000 MW Light Water Reactor thermal nuclear power station at 65% load factor requires 200 tonnes/year of 'yellow cake' (AGR requirements are 10% less).

8.2.2 Fuel enrichment

Reactor grade uranium fuel requires a 2.5-3.5% U_{235} content compared to the natural occurrence of around 0.7%. The process takes place in three stages:-

i. hexafluoride production;

ii. enrichment;

iii. oxide production.

The chemical hazards referred to in Table 8.2.1in the 'hex' and oxide production stages do not present a severe environmental threat in as much as they are likely to be more carefully supervised than many equally dangerous manufacturing processes in chemical plants. Uranium hexafluoride itself has a high sublimation temperature (56°C) so that, if any gas escaped, it would quickly solidify. Nevertheless, great care has to be exercised in ensuring that toxic fluoride and the residual uranium compounds do not find their way into effluent.

---

* Thorium oxide can be used as a breeder fuel in the High Temperature Gas Reactor, although there are no commercial reactors of this type currently planned in the EEC.

Enrichment plants using the diffusion process have the disadvantage of being noisy, large, and, because of their very high electricity requirments, require cooling towers for the waste heat generated. The centrifuge process developed by URENCO (Anglo/Dutch/German enterprise) largely avoids these problems.

The most serious potential hazards during enrichment are posed by the theoretical possibility of criticality during enriched oxide production. Careful designing of plant is necessary to ensure the absence of water from sensitive areas, where it could act as a fission moderator, and to maintain mass of enriched uranium oxide below certain specified levels while it is moving through the oxide production stage.

8.2.3 Fuel fabrication

It will be noted that the only significant risk arising in the fuel fabrication stage, wherein oxide fuel is ground and compressed into thin stainless steel (or zirconium allow) tubes and finally assembled in larger tubes (cladding), is the avoidance of criticality. There is no radiation risk except when reprocessed uranium (or plutonium) oxides are included in the fuel. This undoubtedly will complicate and increase the cost of both the fuel enrichment and fabrication stages.

Although certain Member States are contracted to import a considerable proportion of their enriched fuel requirements over the next 15 years from outside the Community (USSR and USA), in the long term it is probable that most of the Member States' nuclear power industry fuel enrichment/fabrication needs will be met from capacity within the Community.

8.2.4 Irradiated fuel reprocessing

Once nuclear fuel has been irradiated in a reactor, it is still highly radioactive and emits considerable heat. The wastes from a 1000 MWe thermal reactor generate 700 MW soon after leaving the reactor, dropping to about 36 MW after 10 years. It is, therefore, necessary to store the irradiated fuel elements in nuclear ponds at the power stations for some 6 months before they can be moved further, to allow the highly radioactive and short-lived wastes to decay. The irradiated fuel then is made up of the following:-

- U_{238} : approx. 94.4% by weight.
- unburnt U_{235} and Pu that has been produced : approx. 2% by weight.
- high level radioactive but short-lived fission products : approx. 3% by weight.
- long-lived, radioactive, highly toxic actinides : approx. 0.6% by weight.

This waste can either be

i. stored, in the form it leaves the reactor, for 20 years or so;

ii. or reprocessed.

The first option merely delays the second, as eventually some degree of nuclear waste reprocessing is necessary, and there is no doubt that in a large EEC thermal nuclear programme, the building of several fuel reprocessing plants would be unavoidable by the late 1980's. The purpose of irradiated fuel reprocessing is the recovery of uranium and plutonium fuel and the separation of the waste into separate streams so as to facilitate final disposal. (See Section 8.2.5)

The policy towards reprocessing in the U.S. is in contrast to that of most EEC Member States. President Carter has publicly supported the view that the potential economic benefits bestowed by immediately recycling uranium (and possibly plutonium) and the reduction in the volume of high level activity wastes are outweighed by the potential hazards associated with the production of plutonium and the reprocessing operation.

Essentially, fuel reprocessing involves a chemical separation process after the initial outer fuel cladding has been removed. The fuel is dissolved in acid, from which uranium/plutonium is removed by organic solvent; plutonium is chemically separated from uranium and converted to the oxide. The remaining acid solution is reduced by allowing it to boil, and it is this liquor which contains the highly radioactive fission products and actinides.

Based on one year's operation of a 1000 MWe thermal station, the waste streams of reprocessing, besides the separated plutonium, are:-

i. medium to low activity materials from the treatment of storage pond cooling water (which picks up small leakage of fission products from the fuel elements, such as iodine-131);

ii. release of fission product gases, 350,000 Curies* of Krypton-85 and 21,000 Ci of tritium, into the atmosphere;

iii. low level activity solid wastes chiefly from residues of reprocessing low - intermediate liquid waste, 12-15 m^3 of about 45,000 Ci radioactive emission and contains 0.2% of the plutonium produced;

iv. low level liquid waste 250-350 m^3;

---

* Unit of radioactivity (abbreviated Ci). Natural sources of radiation for comparison: tritium in global circulating waters, 100×10^6 Ci; from radium in the oceans, 1000×10^6 Ci; carbon-14 in biosphere (excluding ocean sources), 4×10^6 Ci.

v. fuel cladding waste 3-4 m^3 of about 200,000 Ci, containing about 0.5% of plutonium that was in the irradiated fuel;

vi. highly radioactive liquid waste containing fission products and actinides 10-15 m^3 of about 200 million Curies (after six months cooling);

The recycled plutonium has to be stored in specially constructed vaults under the strictest security.

Before discussing the waste streams, the most noteworthy hazards associated with reprocessing can be identified as:

i. the avoidance of criticality during the uranium and plutonium separation and storage processes;

ii. accidental chemical interaction between acid and organic solvent at high temperature leading to an explosion and release of fission products;

iii. leaks of high activity liquors from cooled storage tanks.

The regular release of fission gases only becomes of concern in a high level nuclear expansion programme. It can, however, be contained. Certain of the low-level liquid wastes are discharged as effluent from the reprocessing plant. Careful monitoring and control is necessary to avoid any unacceptable build-up of radioactivity in the local littoral and marine eco-systems.

The highly active fuel cladding has to be stored in concrete silos where it must remain for at least 15-20 years before some means of final disposal is found.

The high-level radioactive liquid wastes contain mainly fission products, comprising a large number of elements, several of them highly active, so that it is necessary to store these liquors in stainless steel double-walled storage tanks which have to be cooled, and the removal of decay heat. The activity from these fission products decays in 1000 years to less than that of the ore which produced it. This is in marked contrast to the actinides, which are also contained in the high activity wastes. These elements are formed in the reactor from neutron capture by uranium, and their decay daughters. Actinides include elements such as neptunium, plutonium and americum, some of whose isotopes have extremely long decay half-lives, e.g. Ne-237 (2.13 million years), Pu^{239} (24,400 years), Am-243 (7,370 years). They are also highly toxic. There is currently no proven means available to separate the fission products and the actinides, and for subsequent transmutation of the actinides through incineration in a reactor to wastes that would only have to be stored for hundreds instead of thousands of years. And, although research is continuing in this field, many are of the opinion that any process adopted would, itself, introduce new hazards.

Finally, it should be noted, no full-scale oxide reprocessing plant has yet been successfully operated in the world. France is currently operating a 400 tonne/year plant at La Hague, which is to be expanded to 800 tonnes/year by 1982 and intends to build a second 1600 tonne/year plant which will be commissioned over 1985-87.

The UK is currently awaiting the outcome of a planning enquiry to decide whether a full-scale oxide reprocessing plant should be built at Windscale. The experience of two similar plants in the United States, both of which are currently shut down, and the, by no means, incident-free record of the pilot oxide fuel reprocessing plants (which have experienced several shut-downs) leads one to conclude that a testing period of several years will be necessary before it can be demonstrated that such plants can be operated safely and near to design throughput.

In summary, the reprocessing operation, while not essential in the initial years of oxide fuel generation, nor perhaps at all with a small thermal reactor programme (less than 30-40 GW), is unavoidable in the long term. The reprocessing operation itself produces the high level radioactive wastes and separated plutonium, upon which the principal hazards and concern of nuclear power expansion are now centred. Low level gaseous and liquid wastes from normal reprocessing operations should be manageable but careful monitoring and control are necessary.

8.2.5 Ultimate disposal of nuclear waste

On the assumption that reprocessing of the irradiated fuel eventually takes place, the final stage of the fuel cycle is the need for long term disposal of the various waste streams in a manner which will not harm man or his environment. These are best discussed separately.

a. Low level wastes

The disposal of low level liquid wastes from power stations and reprocessing plants presents no threat so long as imposed limits are observed, provided that sufficient dispersion takes place and proper monitoring is undertaken to ensure no unacceptable build-up of radioactivity takes place in marine or littoral eco-systems. Disposal of low level solid waste is carried out either through tipping onto special dumps, storage in caves or underground, or by dumping at sea. Again, there is no cause for major concern over this operation.

b. Intermediate wastes

The method of disposal of intermediate level solid wastes depends upon the type of waste and radiation emitted. Some solid wastes from power stations, which are not gamma emitters, after an initial cooling period, can be disposed of by dumping at sea. Other solid waste requires encapsulating and shielding in concrete or bitumen drums and placed in a specially engineered vault with observation over a considerable period. This particularly applies to the plutonium contaminated fuel cladding. In the UK, the practice to immerse zircaloy hulls from LWR fuel in water, and various methods for decontaminating, separating and disposing of the waste streams from this source are also being considered.

More problematical are the irradiated steel components of reactors which are highly radioactive, mainly because of cobalt-60, which is a gamma emitter. And, although it has a relatively short half-life, and will, therefore, decay to a low activity level over a period of decades, immediate reuse of the site after the retiring of an old thermal station could be prevented unless a satisfactory means of moving and storing this material can be found.

Since the coming into force of the London Convention in September, 1975, the dumping of low and some intermediate wastes in the ocean requires authorisation of the responsible national authorities. The Convention also prohibits the dumping of high level wastes, although there is not total agreement amongst Member States as to what should constitute high-level waste in this context. Most of this dumping is now carried out under the auspices of the OECD's Nuclear Energy Agency who undertake to assess the environmental impact of all dumpings.

c. High level wastes

High level radioactive wastes are initially stored in double walled steel cooling tanks. The method(s) of final disposal of these wastes, because they contain the extremely long-lived radioactive actinides (assuming these are not transmuted to somewhat shorter-lived radionuclides) and certain long-lived fission products (the most conspicuous example being iodine-129 with a half-life of 16 million years), will need to ensure that their radioactive emissions do not enter the biosphere, or, if they do, sufficient dilution takes place as to render the process environmentally acceptable.

Currently, no satisfactory solution for final disposal has been found, although a number of possibilities are under investigation. Because of the potential risks involved with launching the wastes into space, or from ocean dumping, research is being directed towards finding sites in geological formations which are:-

i. geologically highly stable;

ii. impermeable to water leaching;

iii. sufficiently distanced from mineral deposits to avoid any future disturbance from mining;

iv. that have an outside environment interface with good dilution capacity.

The most likely sites are rock formations under the seabed, salt deposits, clay or shale formations and dense crystalline formations such as granite. So far, the most promising of such structures undergoing tests within the Community are the vast salt deposits at Asse in Germany, where there has been no water present for millions of years. Salt also has the advantage of high plasticity which, because of the heat emitted by high level radioactive wastes, would facilitate the natural formation of a compact and stable layer around the waste canisters. Any disposal site will require surveillance and must be accessible for maintenance.

It is also probable that the radioactive liquids will be vitrefied in boro-silicate glass, which stabilises the waste, although at the burial temperatures the glass is not totally proof against leaching. Another problem is the containment of iodine-129, which, if it escaped from the waste canister, would not in the long term be retained in the geological formations, having in mind its lightness and its extremely long half-life.

One is bound to conclude that finding a satisfactory solution(s) for the ultimate disposal of high active waste is likely to take many years. One also has to recognise that even if one solution is found, it may not be accessible to all Member States and its use might be delayed by public reaction. It would, therefore, seem likely that high level wastes will have to be stored in cooling silos for another 10-20 years, and possibly longer still. The risk of radioactive leakage from these silos, although small, is directly related to the volume of vitrefied waste to be stored.

8.2.6 Transport

Containers for the transport of radioactive materials have to conform to standards laid down by the I.A.E.A., and have to be able to withstand a series of impact and fire tests. Providing proper standards are set and observed, we do not feel that this operation presents unacceptable risks. Because of the high energy density of nuclear fuel and its waste, it can reasonably be argued that environmental and safety impact of transporting nuclear fuel is less with nuclear fuel than with other forms of energy. The risks of this operation undoubtedly will be increased with a large FBR programme, with the need to transport recycled plutonium.

8.3 Reactor operation

This section is concerned with pointing out the environmental impacts and risks associated specifically with nuclear power station operation. Those impacts common to all types of thermal electricity generation will be discussed separately in Section 11.

8.3.1 Normal operation

Under normal operation, the pollution and overall environmental impact of thermal and fast breeder nuclear power stations compare very favourably with other forms of electricity generation. Effluent is restricted to low level activity gaseous emissions (tritium,iodine-131/-129 and inert gases) - the level of these emissions varies according to the reactor type, but apart from the possible need to control iodine-129 in a large nuclear programme because of its very long half-life, they can be regarded as harmless; low level liquid wastes - small amounts of very low level radioactive waste from rainwater run-off, floor washings, etc., cooling pond water is recycled; low level solid wastes, e.g. protective clothing and, occasionally, sludge from decontamination of cooling.

8.3.2 Accident risk

Concern over nuclear power stations therefore arises because of the potential consequences in the event of an accident, albeit that the probability of an accident occurring is very low. By its very nature, the fission process in a nuclear reactor operates in a narrow band of criticality; if neutron production is too low, the chain reaction dies; if it is too high, this could lead to a runaway situation. The control process is, therefore, a delicate one. There are also other factors involved in reactor safety which, if not controlled, could lead to release of radioactivity. The most important of these are loss of coolant from failure of process plant and rupture of the fuel elements. Emergency cooling systems are an integral part of all reactor design to control such events.

That control is possible is shown by the generally good safety record* of the industry to date which compares well with other forms of primary energy production.

There are several different reactor designs, the principal features of which are summarised in Table 8.3.1. Excepting the UK, the most favoured designs are the Light Water Reactor systems. Indeed, by 1985, the PWR is likely to represent about two-thirds of total installed nuclear capacity in the EEC. The fast breeder reactor will be discussed separately.

---

*The nearly serious impact of the Three Mile Island Incident can be said to have three results. It highlighted the relative inexperience of operating nuclear plants; it showed that certain hazards, e.g. the formation of hydrogen at very high temperatures from water in contact with zirconium, had not previously been thought of; thirdly it will undoubtedly improve the safety of other nuclear plants and in that some can be said to have lowered the risk of a serious accident.

Table 8.3.1

NUCLEAR REACTOR CHARACTERISTICS

Characteristic \ Reactor Type	MAGNOX	AGR	HTR
Coolant System	Reactor cooled by CO_2 gas. Steam raised in separate exchanger.	Reactor cooled by CO_2 gas. Steam raised in separate exchanger.	Reactor cooled by helium. Steam raised in separate exchanger. Direct use of helium in gasturbine possible.
Efficiency	33%	42%	39% (via steam)
Special Materials	High purity graphite.	High purity graphite.	High purity graphite. Helium.
Fuel Requirement	Natural uranium as metal in Magnox cans.	Slightly enriched (2.3% U-235) uranium oxide in s/steel cans.	Highly enriched (10 to 93% U-235)U/(Th) oxide or carbide in graphite matrix.
Chemical Hazards	Fire risk following air-in-leakage. Some CO_2-graphite reactions.	Some CO_2-graphite reaction. Graphite oxidation after air-in-leakage.	Graphite oxidation following air-in-leakage.
Safety	Low power density so temp. rise would be slow. Loss of coolant by leakage is definite risk. Coolant becomes radioactive. Metal fuel melts easily.	Low power density so temp. rise would be slow. Loss of coolant distinct possibility. Coolant becomes radioactive.	Negative temp. coefficient of activity so intrinsically much safer. Low power density leads to slow temp. rise. Coolant not radioactive. Fuel presents security problem.
Consequences of Accident	Fission products easily released in melt-down. Coolant escape not major hazard.	Fission products not easily released. Coolant escape is not major hazard.	Fission product released very slow. No coolant hazard.
General Comments	Now obsolete - but very reliable.	Many problems in construction. Now in operation, and expected to be cheap and reliable.	Demonstration reactors exist but no real interest at present. Offers high temperature process use potential. Offers Th cycle with breeding possibilities.

Table 8.3.1 - Page 2

NUCLEAR REACTOR CHARACTERISTICS

Characteristic \ Reactor Type	PWR	BWR	CANDU (Pressurised heavy water)
Coolant System	Reactor cooled by water under pressure. Steam raised in separate exchanger.	Reactor cooled by water which boils in core and is used directly in turbines.	Reactor cooled by heavy water under pressure. Steam raised in separate exchanger.
Efficiency	32%	32%	29%
Special Materials	None	None	Heavy water.
Operating Pressure	2235 psia	1050 psia	1285 psia (in coolant)
Fuel Requirement	Slightly enriched (3.2% U-235) uranium oxide in zircaloy cans.	Slightly enriched (2.6% U-235) uranium oxide in zircaloy cans.	Natural or slightly enriched uranium oxide in s/steel cans or zircaloy
Chemical Hazards	None	None	None
Safety	Loss of coolant removes moderator and stops fission - but leaves decay heat. Standby coolant systems necessary. High power density causes rapid temp. rise. Pressure vessel design problems.	As PWR	Moderate power density results in quite slow temp. rise. Loss of coolant definite risk. Pressure tube construction easier and safer than PWR.
Consequences of Accident	Coolant is radioactive so its retention is essential. Fission product release from melting fuel could occur quickly.	As PWR	Coolant retention is necessary. Release of fission products relatively slow.
General Comments	Most popular reactor system by far. Concern over adequacy of pressure vessel safety.	Second-most popular system, but well behind PWR in sales.	Despite high pressures a good design, but requires heavy water availability.

Table 8.3.1 - Page 3

NUCLEAR REACTOR CHARACTERISTICS

Characteristic \ Reactor Type	Organic Cooled HW Moderated	SGHWR (Boiling LW,HW Mod.)	FBR (Liq.Metal Cooled)
Coolant System	Reactor cooled by organic fluid under pressure. Steam raised in separate exchanger.	Reactor cooled by light water boiling in the core, and used directly in steam turbines.	Reactor cooled by liquid sodium. Separate liquid sodium cycle to steam generator as second exchange.
Efficiency	30%	31%	41%
Special Materials	Heavy water organic coolant.	Heavy water.	Sodium metal. Argon as protective blanket.
Operating Pressure	1500 psia (in coolant)	900 psia (in coolant)	20 psia
Fuel Requirement	Slightly enriched (2-3% U-235) uranium oxide in s/steel cans or zircaloy.	Slightly enriched (2.2% U-235) uranium oxide in s/steel cans or zircaloy.	Mixed U and Pu oxides (about 20% Pu) in s/steel cans. Natural U oxide as breeder.
Chemical Hazards	None	None	Fire risk on air-in-leakage. Sodium-water explosive reaction in exchanger.
Safety	As CANDU	As CANDU	Ambient pressure removes rapid loss of coolant risk. Natural cooling circulation prevents decay heat temp. rise after reactor shut-down. Primary coolant very radioactive. Very high power density leads to rapid temp. rise if coolant lost.
Consequences of Accident	As CANDU	As CANDU	Coolant loss would be serious fire hazard. Fuel melt would cause serious activity release including plutonium.
General Comments	Only one reactor plant of this type at present. Offers good Th-cycle breeding potential	As most efficient of the HW types is most likely to develop - but costs seem to be against it at present time.	Very attractive because of breeding potential. Major concern over plutonium system and security.

The risk of serious hazard is a product of the probability of the accident taking place and the possible consequences of that accident. The assessment of the safety risk to the general public of nuclear power station operation is a difficult exercise and one that is not based upon operational experience. Rather, it involves the concept of "maximum credible accident", a technique developed for licensing of reactor designs in the USA and also followed within the Community. Based upon detailed analysis of the operation of the many components of the reactor system, the chances of serious failure of these components is assessed, and an overall probability of "maximum credible accident" is eventually determined. There has been more than one study of nuclear reactor safety, of which the best known are the U.S. 1975 Rasmussen Report (25), the 1976 U.K. Royal Commission on Environmental Pollution - Sixth Report (26) and the U.S. MITRE Nuclear Energy Policy Study Group produced in 1976 (20). The two more recent studies reach similar conclusions.

Both comment that the chances of a runaway reaction leading to a core melt-down and finally a release of radioactive gases and volatile fusion products in sufficient quantity to cause death to the local community are so low that in almost any other circumstances, they would probably be accepted without question. Equally, both comment on the likely public reaction to such an accident, and on the relative lack of reactor operational experience. The actual level of risk of fatality was estimated by these three reports at between 1 in 10^6 and 1 in 10^8 per reactor year*. As further operating experience is gained, it is likely that further improvements in reactor safety will be made. To achieve a further very marked improvement in reactor safety, say by 1 in 10^2, might be very costly and it is, therefore, very difficult to predict how this risk probability estimate might move in the future.

Fast Breeder Reactors

The principal features of Liquid Metal Fast Breeder Reactors, which distinguish it from thermal nuclear reactors are:-

i. that they 'breed' fuel by converting abundant Uranium-238 into Plutonium-239;

ii. the core utilises fast, unmoderated neutrons in the fusion process;

iii. because of the need to transfer heat more rapidly from the highly compact core and avoid reactor slow-down, the fast breeder uses liquid sodium as coolant (inert gases can be used):

iv. unlike the PWR, loss of coolant does not lead to a slowing down of the chain reaction.

---

* This compares with the risk of fatality from natural disasters for an individual per year of 1 in 500,000 (Rasmussen Report and Flowers Report). Incorporated in this probability is a 1 in 10^2-10^3 chance of the weather conditions and the location of population being such as to expose local inhabitants to severe radiation. The application and use of the Rasmussen results have been criticised in the United States and these risk estimates should be read in the light of footnote comments on Page 90, discussing the implications of Three Mile Island.

From a long term economic point of view, the attraction of the FBR is that eventually it will not require fresh uranium supplies to be made available. However, it should be pointed out that independence of nuclear fuel supplies can only be achieved if there is prior commitment to a substantial thermal nuclear programme. From an environmental point of view, its disadvantages are that it produces plutonium, which is potentially the best nuclear bomb material as well as being highly toxic. Secondly, in the event of a runaway incident and a core melt-down, the containment of radioactivity would be more difficult because of the intrinsically faster nuclear reaction taking place. Thirdly, as well as being radioactive, the liquid sodium coolant is highly chemically active, so that again the containment problem is of paramount importance.

A particular concern pertaining to FBR's has been over the sequence of events that could follow from boiling of the liquid sodium adjacent to the fuel element if, for some reason, there was restricted coolant flow. It has been suggested that 'voiding' could lead to an explosion or a very rapid rise in reactivity with the result that the fuel elements might become sufficiently ruptured or distorted to lead to a runaway reaction. However, a recent experiment conducted on the UK Dounreay EFR reactor (before being shut down) would seem to go a considerable way towards allaying many of these fears. The experiment involved the deliberate blocking of the coolant channel (90% restricted flow) adjacent to one of the fuel elements for over 24 hours to observe the results. A full analysis of the results are to be given in two papers (27, 28) by C.V. Gregory and D.C.C. Smith later this year. The experiment showed that not only do the instruments give rapid response to the incident, but, in fact, no violent energy release occurred and none of the other fuel elements were damaged. This would indicate that the liquid metal coolant with its high heat conductivity contributes to a stable reactor system under severe conditions.

8.4 Nuclear proliferation and socio-political impact

Both these issues arise because it is practically possible to assemble a nuclear device from 10-15 kg of fissile material containing isotopes U^{235}, Pu^{239} and U^{233}, the yield and effectiveness of the device depending among other things on the proportions of other isotopes. Theft of this material from different stages of the nuclear fuel cycle could create the possibility of two events taking place:-

a. nuclear proliferation

a non-nuclear power could manufacture a bomb, either through enrichment of stolen UO_2 and burning the fuel in an experimental reactor and subsequent processing of the irradiated fuel, or by processing of stolen reprocessed fuel;

b. <u>terrorist threat</u>

an extremist group could, in principle, assemble a crude device by separating plutonium in a secret processing laboratory and use it to threaten society unless certain demands were met.

It is these possibilities which are the chief concerns of many nuclear power opponents.

Another argument that is raised, which arises directly out of the possibility of these events, is that their prevention could only be safely ensured by a secret police force whose powers, in order to be effective, would have to include various forms of surveillance and rights of detention. As such, their presence would represent an infringement of civil liberties. This argument is not conceded by the nuclear power industry who claim that prevention of theft can be achieved through technological means and suitable armed guard of the nuclear fuel in transit.

The two potential national and international security threats need separate consideration since the possibility of their occurring would differ according to the form and extent of the nuclear power programme embarked upon in the Community.

8.4.1 Nuclear proliferation

The possibility that nations without nuclear weapons could acquire the means to produce a nuclear deterrent from the presence of a nuclear power programme in the Community could occur

i. by theft of natural uranium being imported to the Community, which would subsequently have to be enriched in facilities (likely to include an experimental reactor) in the country concerned, or by another country in secret alliance;

ii. by theft of fissile material from fuel fabrication plant, nuclear power station, or from fuel processing plant;

iii. hijacking of irradiated fuel moving between the plants mentioned in (ii).

The threat of natural uranium oxide theft is considered small relative to (iii) because of the fact that the nation concerned would also have to have the means to enrich and produce the fuel for subsequent irradiation in a small reactor in order to obtain weapon grade plutonium U^{235}/U^{233}. Nevertheless, in 1967, 200 tonnes of uranium oxide ('yellowcake') disappeared while en route for France. The potential theft from power plants themselves must also be considered very low given the high level of supervision in sensitive areas of power stations, and the ability to design virtually automatic fuel handling systems with a high degree of sensitivity. Also, it is extremely unlikely that hijacking of highly active irradiated fuel in transit from power stations will take place because of the risk and difficulties in handling this material.

8.4.2 Terrorist threat

Given the relative lack of resources available to terrorists compared with nations (even very undeveloped countries), it is probable that only option (iii) and possibly theft from reprocessing plant would be considered by a group intent on acquiring the means to manufacture a bomb. In other words, the terrorist threat is considerably reduced if no reprocessing plants are built, i.e. if nuclear waste is all stored in nuclear ponds. It has been argued that the same objective could be achieved by building nuclear parks which would locate very large nuclear power station capacities in the same complex as enrichment/reprocessing plant. However, the land use and other potential environmental impacts and risk are also considerably heightened by such developments.

It should be noted that enriched reactor grade fuel can also be detonated (ERDA* reported on September 14, 1977, that the United States had recently exploded a reactor grade fuel), but this is much less likely to be attractive to would-be hijackers.

Because plutonium represents the most convenient potential bomb material, these social and political risks are increased by a faster breeder programme with its inherent need to separate and move plutonium. The most sensitive operations would be fuel enrichment, as well as fuel reprocessing, and the transport of the material between these stages. Again, it is probable that the risks involved with hijacking of the irradiated fuel would deter terrorist activity at this stage of the cycle.

---

* The U.S. Energy Research and Development Agency, now within the Department of Energy.

9. IMPACT OF FOSSIL FUEL PRODUCTION AND USE

The use of fossil fuels in meeting energy needs has associated with it a number of potential impacts upon the environment. From Part A of the report, it can be seen that there is a considerable difference between the level of fossil fuel production and consumption in the Community. This section on fossil fuels will begin by discussing the impacts associated with coal and hydrocarbon production within the Community and then go on to examine the several different areas of potential environmental damage from fossil fuel consumption, and the scope for reducing this impact.

PRODUCTION

9.1 Coal

The extent and type of environmental impacts of coal production depend very much on the method of mining employed. There are two broad classifications - deep mining and open-cast extraction. Lignite production (mostly confined to Germany within the EEC) is entirely based upon open-cast mining.

9.1.1 Amenity/social disturbance

In both methods of extraction, considerable local disturbance and loss of amenity is involved with opening up a new pit. This is obviously more marked with open-cast coal mining, and, in Germany, this has even resulted in the moving of whole villages. With modern deep mining techniques, the disturbance and the loss of amenity involved in opening a new mine is less than those of traditional methods and considerably less than those associated with open-cast mining. Even so, it can be expected there will often be fierce local opposition to opening up new deep-mined coal pits, particularly in rural areas which do not have a history of coal mining.

The loss of land involved in modern deep-mined coal production can range from 12-20 hectares per million tonnes hard coal production. For an open-cast pit, approximately 200-500 hectares per million tonnes hard coal production is required (in the case of lignite production, this figure is roughly quadrupled).

The higher level of disturbance and loss of amenity in open-cast mining is partially compensated for by the fact that, with modern land reclamation techniques, the spoil and over-burden can be replaced and landscaped within 5-10 years of opening up production.

9.1.2 Noise and particulates

Facilities for handling coal, whether it be indigenously produced or imported, are inevitably somewhat noisy and can produce coal dust. While considerable reduction of impact has been achieved in more modern equipment, it will normally be necessary to locate bulk coal terminals away from residential areas.

9.1.3 Solid waste

Colliery spoil arises from the sedimentary rocks and shale with which seams of coal are deposited. The proportion of waste can vary between 15% and 60% of the volume of coal or lignite extracted, and depends upon the thickness of the coal seams and the method of coal extraction. It is estimated that the average solid waste production of deep-mined coal in the EEC (29) is around 0.5 tonnes of colliery spoil per tonne of coal/lignite mined. However, it is expected that this figure will gradually decrease as more modern mining techniques are used, and because many of the new fields discussed in the UK have very thick seams.

With open-cast mining, the spoil is used as back-fill in reclaiming the land. A small proportion of the total waste can also be used as land-fill and in road construction and in manufacturing building products (about 10-15%). Nevertheless, in many areas of high mining and coal combustion activity *, the solid waste is likely to create considerable problems of disposal.

The potential danger of water pollution of local streams and rivers by acid run-off from tips can largely be contained by modern methods of compacting, landscaping and, if necessary, directing the water through a treating plant.

9.2 Oil

The most serious potential impacts of oil production relate to offshore activities. Under normal operation there is bound to be some interference with the fishing industry. It is not yet possible to say whether the accumulation of small routine spills from offshore operations is likely to have a detrimental impact upon the northern North Sea marine environment. The presence of platforms themselves could well tend to conserve fish populations.

The real threat from offshore oil operations is the damage that could result from an accidental major oil spill. In spite of some improved operational practices being introduced following the 1977 Ekofisk blow-out, there are inevitably going to be further major spills resulting

---

* The two often go together so as to minimise transport costs.

from expanded North Sea oil production. The Norwegian Ministry of the Environment has recently estimated that in the worst circumstances a future North Sea blow-out could lead to a spoil nearly ten times as large as that from the Ekofisk incident (about 20,000 tonnes over 7½ days (30)). If an underwater crude oil pipeline were fractured, the resulting spill could be as much as 2-3,000 tonnes/hour of oil.

However, it would appear that the prevailing wind, the sea conditions and tidal drift in the North Sea are likely to considerably ameliorate the potential impact of such accidents. No detectable quantities of oil reached the Norwegian shore from the Ekofisk blow-out (30) in spite of the inability of the oil clearing equipment to operate efficiently in the size of waves prevailing at the time. The potential impact of an oil spill on the marine environment will to some extent depend upon its timing with respect to fish and plankton breeding cycles. In fact, soon after the Ekofisk incident, the Norwegian Institute of Marine Research and the UK Ministry of Agriculture and Fisheries found, after conducting a series of tests on plankton just beneath the surface of the sea area affected, that little damage had occurred.

Natural gas

The only serious environmental effect of natural gas production and importation is the risk of an explosion in the gas plants, storage tanks (when the latter are empty) or LNG/LPG carriers themselves, so that it is necessary that gas plants and terminals should be sited well away from residential areas and other industrial activities.

COMBUSTION IMPACTS - GLOBAL

Impact of carbon dioxide (CO_2) emissions on climate

CO_2 is produced as the final oxidation product of carbon dioxide in fossil fuels. There are no practical means of controlling this emission. It is estimated that 15×10^9 tonnes/year are produced from world fossil fuel consumption, of which just over 3×10^9 tonnes/year is emitted from the EEC. This should be set against an enormous natural cycle of CO_2 production and absorption. Plant photosynthesis absorbs 150×10^9 tonnes/year (31) and a corresponding amount is returned to the atmosphere by decay (32). The land masses and oceans also act as natural sinks for carbon dioxide. The atmospheric mass of CO_2 is 2.4×10^{12} tonnes.

The concern with respect to man-made CO_2 emissions arises over the possibility that an increase in CO_2 concentration in the atmosphere over time might result in an increase in ambient air temperatures which in turn could result in significant climatic change. This change is caused by the 'greenhouse effect' which is a global warming process caused by increased absorption in the atmosphere of re-radiated solar energy from the earth. The current state of knowledge about this process and related feedback mechanisms is very limited. Very approximate measurements of atmospheric CO_2 concentration in Hawaii have indicated that global atmospheric CO_2 concentrations may be increasing by some 0.2-0.3% per annum (33).

On present evidence it is impossible to conclude whether or not the increase in CO_2 emissions from the combustion of fossil fuels is resulting in a rise in global temperatures.

9.5 Particulate emissions

Atmospheric particulates have a confirmed role in the determination of weather in, for example, cloud formation and scattering and absorbing solar radiation. This factor has prompted a number of observers (35) to postulate that particulate pollution may be a modifier of climate, acting to depress world air temperature. Again, there is inadequate data available to evaluate the risk. Nevertheless, it is worth noting that while current particulate abatement technology is relatively advanced and affords very substantial emission reductions, the small proportion of matter which evades capture consists predominantly of fine, non-reactive, non-hydroscopic particles which are very long-lived in the atmosphere and, hence, are most likely to give rise to the effect postulated.

LOCAL AND REGIONAL IMPACTS OF COMBUSTION

9.6 Sulphur oxides (SO_x)

The release of sulphur oxides (SO_2 and SO_3) into the atmosphere in Europe is very largely (over 90%) due to the activities of man, of which the burning of fossil fuels constitutes the major source. An overwhelming proportion of sulphur oxides emitted is in the form of sulphur dioxide (SO_2) and it has been estimated that currently some 70% of total SO_2 emissions in the Community originate from power stations.

The existence of health hazards and environmental damage from SO_x pollution has been known for many years, and the effects have been studied extensively in Europe and in the rest of the world. The most recent review of the health hazards from combustion of fossil fuels has been carried out by the Umweltbundesamt (within the German Ministry of the Interior) (49). Correlations between the extent of health damage and the level of pollution have been clearly demonstrated, although there is disagreement about the linearity of these relationships and whether or not threshold levels exist, below which there is no effect.

Concern about the environmental impact of SO_x emissions is focussed primarily on the following effects:-

- at sufficient concentrations, SO_x in the gaseous phase can lead to chronic injury or mortality of vegetation;

- physiological damage in man and animals can result in the form of respiratory and cardiovascular aggravation, bronchitis and eye irritations; this had led to deaths, principally among old people;

- SO_2 and SO_3 can combine with atmospheric moisture to form H_2SO_4, leading to the phenomenon of 'acid' rainfall and subsequent impacts in the form of corrosion damage to many materials, damage to vegetation, and acidification of surface waters;

- the potential occurrence of complex interactions with other pollutants present in the atmosphere leading to the enhancement of pollution impact, the most notable being the formation of sulphate particulates. As well as their health hazard, sulphate particulates result in reduction of visibility.

9.6.1 Local impact

Most SO_x emissions are of a highly localised, point source nature. Of prime importance are the first-order, human and biological impacts associated with high, local concentrations of SO_x at or near ground level (especially in urban air). The technique most widely employed in the Community for preventing build-up of high SO_x concentrations is the tall stack, where emission dispersal is the aim. There is considerable evidence to support the view that tall stacks are an effective (and low cost) method of preventing high, local, and in some instances, regional ground level concentrations of SO_x.

As will be discussed a little more fully below, the relationship between SO_x emission and deposition of sulphur oxide pollution in one form or another is complex. The most important factors influencing the relationship are:-

i. concentration of source emissions;

ii. local topographical features;

iii. climatic conditions.

These factors will determine the effectiveness of high stacks in alleviating local and sub-regional ground level concentrations of sulphur oxide pollution.

9.6.2 Regional/global impact

Evidence is being accumulated which suggests that point source emissions of SO_x are transported over long distances to areas remote from the source which may then lead to more subtle, long term environmental damage, particularly to vegetation and aquatic eco-systems. The most recent and extensive report on this subject was conducted by OECD (35). This study investigated the long range transport of air pollution and concluded that sulphur compounds do travel long distances in the atmosphere and that the air quality in any one European country is measurably affected by emissions from other European countries. A summary of data, developed during the OECD study on emissions and receipts of atmospheric sulphur amongst participating countries is contained in Table 9.6.1 overleaf. An examination of the data in Table 9.6.1. indicates that the Community is a substantial net exporter of SO_x (20-60% of SO_x emitted), principally to Scandinavia and other European countries. Import of SO_x would only seem to account for about 5% of SO_x deposition in the Community. The findings of the OECD study imply that if some European countries find it desirable to reduce substantially the total deposition of sulphur within their borders, individual national control programmes can achieve only a limited improvement while, at the same time, national SO_x abatement programmes will result in reductions in sulphur pollution experiences in other countries.

The incidence and degree of eventual precipitation of sulphates/SO_2 exported from the Community is also considerably dependent upon the prevailing weather conditions, and consequently can be highly episodic in nature.

9.6.3 Methods of control

There are several methods currently or potentially available for reducing SO_x emissions. These can be generally classified as follows:-

- burning of fuels having a naturally low sulphur content;
- desulphurisation of the fuels;
- desulphurisation of flue gases;
- fluidised bed combustion.

Table 9.6.1 Estimated budget for dry plus wet deposition of sulphur for 1974. Unit: 10^3 tonnes S. Numbers rounded to one significant figure and accurate to within about $\pm 50\%$. The sums are calculated from unrounded figures and thereafter rounded separately.

EMITTERS / RECEIVERS	Austria	Belgium	Denmark	Federal Republic of Germany	Finland	France	The Netherlands	Norway	Sweden	Switzerland	United Kingdom	Czechoslovakia	German Democratic Republic	Italy	Poland	Other Areas	Undecided	Total received from all areas	Total emitted from all areas
AUSTRIA	60	6	0	40	0	20	2	0	0	5	20	20	20	30	7	20	30	300	221
BELGIUM	0	100	0	20	0	30	5	0	0	0	30	1	4	0	0	1	10	200	499
DENMARK	0	1	60	6	0	3	1	0	2	0	10	1	6	0	2	2	10	100	312
FEDERAL REPUBLIC OF GERMANY	8	60	7	700	0	100	40	0	2	7	100	20	80	7	10	10	90	1300	1964
FINLAND	0	2	8	10	100	4	2	2	30	0	10	7	30	0	20	80	70	400	274
FRANCE	2	40	1	50	0	600	10	0	0	6	100	5	20	30	2	30	150	1000	1616
THE NETHERLANDS	0	10	1	10	0	10	60	0	0	0	30	1	4	0	1	0	10	150	391
NORWAY	0	4	8	10	1	9	4	30	9	0	60	3	10	0	5	4	100	250	91
SWEDEN	0	7	30	30	10	10	6	6	100	0	40	8	50	0	20	30	100	500	415
SWITZERLAND	1	2	0	7	0	20	1	0	0	30	10	2	1	6	1	2	20	100	76
UNITED KINGDOM	0	8	2	10	0	20	4	0	0	0	800	2	9	0	2	1	100	1000	2883*
CZECHOSLOVAKIA GERMAN DEMOCRATIC REPUBLIC, ITALY, POLAND AND OTHER AREAS	60	60	80	400	40	200	40	9	50	10	600	900	1300	900	1000	4500	1000	11000	-
SUM	100	300	200	1300	150	1000	200	40	200	60	1800	1000	1500	1000	1100	4600	1900	17000	-

* Including 80.10^3 tonnes S from Ireland.

SOURCE : OECD Programme on Long Range Transport of Air Pollutants. Measurements and Findings, Paris, 1977.

The fuels which are the principal sources of sulphur emissions in the Community are:

- coal : sulphur content approximately 1-2% wt.
- fuel oil : sulphur content approximately 1-4% wt.
- distillate oil : sulphur content approximately 0.5-1.0% wt.

Sulphur emissions in the Community have been significantly reduced over the past twenty years, principally by the substitution of low sulphur natural gas and gas oil for coal, but also by the use of lower sulphur content crude oils, supplemented by distillate oil desulphurisation. Desulphurisation of fuel oil has been achieved entirely by indirect methods, involving the blending back of desulphurised distillate fractions. Direct hydro-desulphurisation of residual fuel oil is a high cost process ($10-30* per % wt. sulphur/tonne), and is uneconomic given the current market price differential ($7-9/tonne) between 1% and 3.5% sulphur fuel oil. Increased demand for low sulphur fuel oil (e.g. stemming from stricter emission standards) would be expected to increase this differential and might eventually make direct fuel oil desulphurisation economically viable. This is unlikely to occur before 1985-1990 in view of the increase in availability of low sulphur North Sea crude oil.

Direct desulphurisation of coal is very much in the experimental stage, although considerable progress is now being made towards this end in the United States (36). Flue gas desulphurisation after combustion is thus the only current means of controlling SO_2 emissions from coal.

Very little flue gas desulphurisation is currently taking place in the EEC. Various techniques can be used, but that most widely adopted in Japan and the United States is that of lime and limestone scrubbing. There is still a certain amount of unreliability attached to these processes and they are costly. For a coal-fired power station, limestone scrubbing (reduces sulphur emissions by about 90%) is estimated to increase the cost of generating electricity by about 20-30% (37,20)**. Further development is likely to reduce this cost. Limestone scrubbing creates an additional waste disposal problem, namely that of dealing with the wet sludge produced (0.05-0.2 tonnes of sludge - 60% solid, 40% liquid - is produced for each tonne of coal burned assuming 0.5%-2.0% sulphur content respectively).

---

* Cost depends on actual sulphur content of fuel oil.

** The range is partly dependent upon load factor of plant.

The development of fluidised bed combustion is, in our view, an important one for the Community, as it offers the prospect of substantial reductions in emissions of SO_x and other air pollutants at relatively low costs compared to flue gas desulphurisation. Part of the lowering of cost for industrial boilers comes about because of improved combustion efficiency. This system is nearing commercial development for industrial application (perhaps within the next four years), while application for electricity generation is somewhat more remote (maybe 10-15 years).

9.7 Other atmospheric pollutants (NO_x , particulates, CO, HC)

Of the other fossil fuel-associated air pollutants, nitrogen oxides and particulates are generally regarded as being of greatest importance for the environment. Nitrogen oxides (NO_x) at high concentrations are toxic to man and most animal species and cause biological damage in many other organisms (38). NO_x is known to play an important role in the formation of photochemical pollution in urban atmospheres, where technological emission sources are of greatest significance (particularly from motor vehicles and power stations). OECD (35) also estimated that about 25% of 'acid rain' precipitation was formed from nitrogen oxides, and that the proportion was sometimes as high as 50%. Like sulphur oxides, NO_x are also transported long distances (35).

Techniques for controlling NO_x emissions from both mobile and stationary sources aim to prevent formation by modifying combustion conditions or to remove NO_x from the combustion gases after formation. With stationary sources, although operating experience is very limited, combining various combustion control techniques appears to be particularly promising, and may eventually, in some instances, offer overall savings in cost (39). Flue gas NO_x removal is a more remote possibility, though processes employing aqueous scrubbing for combined control of SO_x and NO_x seem to be technically feasible (40). Our earlier remarks on flue gas desulphurisation costs also apply here, and these are likely to be substantially higher than those for combustion modification. Our views on the importance of fluidised bed combustion are equally relevant to NO_x emissions from stationary sources.

With regard to NO_x emissions from mobile sources, various processes are still in the research stage. The use of catalytic conversion units is, however, already being employed outside of the Community. The two main obstacles to the use of this type of process are that it is presently unreliable in terms of sustained performance and that it would require lead-free gasoline to be widely available. This last point raises a number of problems but probably the most important is that removing lead additives from gasoline would increase manufacturing costs by between 15-40% (41,42).

Apart from the potential global impact of combustion-generated particulate pollution discussed earlier, this class of pollutant is important for local environmental impact for the following reasons:-

- at high concentrations, it can cause interference with respiratory functions, eye irritations, etc.;
- it can result in damage to materials, soil and water contamination, damage to vegetation, paint discolouration, etc.;
- it reduces local visibility;
- fine particles can act as catalytic surfaces in the atmosphere for other chemical reactions which can lead to synergistic pollution effects, e.g. enhancement of SO_2 impacts;
- it reduces local solar influx and may cause local interference with weather.

Our earlier comments on particulate control also apply here, though we stress again that, while overall emissions of particulates have been, and are likely to continue to be, drastically reduced , the nature of the remaining particulate emissions is such that environmental impact is unlikely to be reduced in direct proportion.

Carbon monoxide and hydrocarbon emissions in urban areas arise predominantly because of the incomplete combustion of fossil fuels. In the case of carbon monoxide, the actual emission levels of CO are the highest of all atmospheric pollutants (20). However, their background level has not increased because of the many natural sinks for this product. The principal concern therefore relates to its potential toxicological effects on humans in high urban concentrations, particularly from motor cars. The principal source of unburnt carcinogenic hydrocarbons to the public is also motor vehicles. As combustion efficiency improves, it is reasonable to expect that the potential health hazards of these pollutants should be contained.

9.8 Water pollution

The main areas where the production and use of fossil fuels have some consequency for water pollution are:-

- effluents from coal mining (particularly acid mine drainage and spoil run-off);
- effluents from coal beneficiation and conversion;

- discharges (particularly accidental release) of crude oil to the marine environment from offshore production and transportation operations;

- effluents from oil-refining;

- discharges of thermally-charged cooling water from power stations.

Given the current state of technology, we do not believe water pollution to be a major issue in relation to the future processing and use of fossil fuels in the Community. The categories of effluent listed above are very largely treatable to acceptable standards with present techniques. Some local disturbance of surface waters is almost bound to occur from coal mining, but a very great deal can be done to minimise the impact. Water pollution from coal processing and conversion can be reduced virtually to zero without serious difficulty or unacceptable cost , and we have assumed that such treatment and effluent recycling will be employed in all future coal gasification plants in the Community. Effluents from oil refining are already extensively treated and recycled in the Community and further improvement is achievable to the point where impact would be negligible.

9.9 Solid wastes

Solid wastes generated during the use of fossil fuels arise primarily from:-

- residual solids from coal processing and conversion;

- fly-ash and clinker from coal-fired plants.

Residues from coal processing plants are of a very similar nature to colliery spoil. Again, we anticipate that future plants will have to apply all practicable environmental controls. It is likely that some of the future coal gasification plants will be located in the vicinity of the supplying mine which may permit the return of solid and sludge wastes to the mine for disposal. Alternatively, restoration techniques should ensure that any associated impact of disposing of coal processing wastes is of a confined and temporary nature.

Solid wastes from coal combustion are very different to other types of coal-associated waste, being for the most part inert and relatively stable. Their unique properties enable them to be used as by-products for a wide range of purposes (cement production, brick making, etc.). Although substantial quantities are already used in this way in certain Member States, the opportunities for disposal by this route are inherently limited. Landfilling of coal-combustion wastes involves virtually no risk of pollution damage and, indeed, may be environmentally beneficial in reclaiming derelict land. However, we see a potential problem in future of disposing of these wastes in areas where suitable disposal sites are not available in proximity to the plants concerned as, in the quantities in which these wastes arise, transportation of the wastes over long distances is prohibitively expensive. This problem is likely to be particularly acute for coal-fired plants in urban areas, and in coal mining areas.

The problems of solid waste handling and disposal are exacerbated by the introduction of SO_x emission control techniques. With limestone scrubbing, a wet sludge is produced which is partly dehydrolysed on site but finally has to be disposed of in sludge form containing about 3 times the weight of fly-ash. With lime fluidised bed combustion, the overall volume of solid waste is only increased by 25-30%.

9.10 Land requirements

Facilities for the extraction, processing and use of fossil fuels require substantial areas of land. This is particularly so in the case of:-

- coal mines : deep mined (30-150 hectares*)
  open-cast (150-250 hectares/million tonnes produced**);

- coal processing and conversion plants (50-200 hectares);

- oil refineries (150-200 hectares).

Clearly, the requirement for land for coal mining is restricted to areas where coal deposits are located. In some cases, this may involve removing the land from other high value uses (e.g. intensive agriculture). Broadly speaking, the same applies to coal processing plants and power generating stations, where these may, in future, be competing heavily with other desirable uses for suitable sites. The tendency for oil refineries to be located on the coast has, in the past, mean the removal of some prime coastal sites from alternative use, in particular recreation.

9.11 Demand for water

Substantial requirements are generated by coal conversion plants and, to a lesser extent, by oil refineries. Demand for water by thermal power plants will be discussed in a later section.

---

* Large coal fields can have up to five or six pit head sites.

** The total area finally excavated from a single open-cast working may be in excess of this figure, but back-filling of spoil and overburden during mining should contain the total land required for open-cast mining at any one time.

In the case of coal gasification plants, demand for water originates both from the water consumed in the process of manufacturing synthetic natural gas and from the need to make up evaporative losses and purging water for cooling purposes (recirculating cooling). By maximising air cooling (this can increase plant operating costs), water requirements can be reduced by 5-10%. Even so, a fullscale coal gasification (high BTU) plant (8.5 million tonnes/year coal feed) will require about 34.5×10^6 metres^3 of process water, and a further 40×10^6 m^3 for use in assisted cooling of the waste heat. It is likely that coal SNG plants will tend to be located near to coal fields (the gas will be distributed via existing natural gas trunk lines). The demand for water from the plants could,therefore, be concentrated in specific regions of the Community and could represent a substantial increase in total water requirements of the area.

10. IMPACT OF RENEWABLE PRIMARY ENERGY DEVELOPMENT

The development of 'renewables' is often most ardently advocated by those who oppose the expansion of nuclear power on the grounds of the latter's potential harm to the environment, or because of the increased possibility of nuclear weapons proliferation and terrorist threats posed by the existence of the operation of the nuclear fuel cycle. However, what are seen as alternative energy sources to nuclear power are not without certain social and environmental impacts of their own. In most cases, this impact is local rather than regional or global in nature and, generally speaking, it is unlikely that their operation will present a health hazard.

10.1 Solar energy

Its various possible forms of development need to be treated separately.

10.1.1 Heating panels

The only long term environmental impacts that are likely to result from the maximum utilisation of solar energy for domestic/commercial heating are both visual. New buildings will need to have the correct orientation to the sun. This could constrain the location and layout of new development, as well as the design of buildings themselves. Design will be particularly influenced by the configurations imposed by maximum passive solar energy utilisation.

10.1.2 Heliostat and photovoltaic electric power generation

Though unlikely to be very significant in EEC countries, banks of mirrors or silicon solar cells would require several times the land area of that required by a thermal power station to generate an equivalent quantity of electricity.

10.2 Wind power

The potential visual impact of rows of 2-5 MW wind generators, each being anywhere between 25 and 50 metres high, would be very considerable if they were to be constructed on hilltops or along sea coasts. It is probable that any large-scale development of this kind would run up against considerable resistance from local and amenity lobby groups. If they were to be constructed offshore, this problem would be avoided, but at considerable cost. There would be a certain amount of noise generated by giant wind turbines, and their land use requirements would be approximately ten times that of an equivalent capacity thermal power station. The building of very many overhead electricity transmission lines would also present a considerable visual intrusion.

10.3 Tidal power

The effect of a large barrier across a tidal estuary utilising the tidal flow to drive turbines would be to narrow the rise and fall of the tide upstream of the barriers. This would affect the water table of the area and have a pronounced impact on the flora and fauna of the littoral zone and possibly also on the inland region. Also, since water flow would be much reduced, there would be an effect on marine life. The precise nature of the impacts could only be determined through a careful study of the various eco-systems of the area likely to be affected. The preliminary feeling of most experts who have studied the Severn estuary in the U.K., is that the impact of a barrier would not cause large environmental damage, and it is possible that some effects could be beneficial. Economic and social life in the region of the barrier would also be affected (43,44).

10.4 Wave power

To achieve the equivalent output of a 1200 MW thermal power station, a string of wave machines would have to be built out to sea some 4 kilometres long. Thus, if any large-scale development of wave power is to take place, considerable stretches of ocean will have to be marked off and lit so as to avoid being hit by shipping. The size of the offshore structures necessary to support the wave machines is likely to be considerable, particularly those platforms which might house generating/transforming equipment, chemical means for storing electricity and operating staff. Depending on the exact location for wave generators, it is possible that the activities of the fishing industry could also, to some extent, be restricted.

A reduction of wave motion in the littoral zone is also likely to have an impact upon the flora and fauna of the coasts affected.

The most likely areas for situation wave power machines are off the north west coast of Scotland, the Western Approaches to the Channel and off the west coast of Ireland. The Western Approaches location suffers because of the amount of sea traffic. The main objections to the Scottish location will be the impact of the transmission lines.

10.5 Biomass

The impact of biomass collection, in whatever form it takes place, would be very considerable on flora and fauna of the areas affected, and the noise and visual impact of the collection and transportation operations would probably also be marked. Because of their low energy density compared to fossil fuels, the preparation and combustion of biomass fuel would also have considerable demand on land, water and nutrient requirements. The utilisation of waste-derived fuel would not carry most of these negative environmental impacts.

10.6 Combined heat and power

The most significant impact of CHP development is likely to be the disruption caused during the lengthy time (up to 15 years) it takes to lay down and bury the lagged pipelines during the building of the district heating system. It is also likely to be most successful in high density housing conurbations, which presumably would have to accept the presence of nearby power stations as part of the visual landscape. Once laid-on, district heating systems tend to limit the choice of fuel that can be exercised by the consumer, but the point has been made that this is already the case with much public housing and for tenants in rented accommodation.

10.7 Geothermal

Utilisation of geothermal energy through vapour systems creates considerably less environmental problems than those based on hot water. Even so, the steam, after passing through the turbines, may be highly mineralised. The quantity of water is about three times that of the flow through a directly cooled thermal power station of equivalent capacity (45). Reinjection of water may, therefore, be necessary. In hot rock hydrothermal schemes, circulating water will be essential to the operation anyway.

11. IMPACT OF THERMAL ELECTRICITY GENERATION

The provision of additional electricity generating and transmission capacity has associated with it several pronounced environmental impacts. This section is concerned with describing those impacts which are common to all forms of thermal generation. Those impacts deriving from the particular fuel type being burnt, whether it be nuclear or fossil fuels, have already been described. Apart from the proportionately high degree of thermal pollution emanating from such power stations, the potential impact of which was discussed in Section 7, other forms of environmental impact from this source are principally local in type, although there could be regional implications.

11.1 Impact on amenity

Because of their enormous size, it is hardly necessary to point out the disturbance to visual amenity caused by the building of a modern thermal power station of 1300-2000 MW. It is probable that most new capacity will be located in relatively rural environments. While good layout design, architecture and landscaping can do much to improve the appearance of such colossal structures, considerable visual intrusion will usually be unavoidable. The size and appearance of thermal power stations is markedly affected by whether or not it is directly cooled by circulating water from a nearby large source of water such as a river, estuary or the sea. In this case, there is no requirement to build the giant cooling towers. However, acceptable new river or estuarial sites for thermal power stations will, in many Member States, be increasingly hard to find. Coastal sites, because of their generally higher amenity value than inland locations, will also be increasingly less easy to come by.

It is, therefore, expected that the majority of new power station sites will be inland and will require assisted cooling, utilising a recirculating, evaporative system, which transfers the heat of the water to the air by means of large cooling towers. The number of towers can vary from two to six, and are 100-150 metres in height. The variation partially depends upon the capacity of the station, but also upon whether higher cost mechanical draught-assisted systems are employed.

Considerable impact on the rural environment will also result from the need to build high-voltage transmission lines connecting new power stations to the super-grid. These connecting links may be anywhere between 10 and 60 kilometres in length. An increase in base-load generating capacity is also likely to lead to a similar increase in super-grid transmission capacity, as well as additional associated switching and transforming stations. The burying of high voltage transmission lines costs about 15 times as much as the building of overhead pylons. Such economic penalties are likely to mean that only a small proportion of new high-voltage transmission lines is buried.

Because of the pronounced visual impact of both thermal power stations and overhead high-voltage transmission lines, it is to be expected that, for many sites, there will be considerable opposition encountered to construction of such projects. The building of thermal power stations takes from 6-8 years to complete and employs a peak temporary workforce of around 2,500-3,000. In more isolated areas, this can obviously have a marked socio-economic impact on the local community.

11.2 Water/thermal pollution

"Once through" direct cooling systems takes either fresh or salt water from a local source through the boiler/reactor of the power station and returns it to the source. A typical 1000 MWe thermal power station heats about 30 m^3 some 10^oC per second. The effect of such thermal pollution depends on the eco-system and size of body of the water being used as the source. Sites for cooling water discharge have to be selected where sufficient dilution of the thermal pollution can take place so as to avoid adverse environmental impacts.

To the extent that there are dissolved, inorganic solids or acid in the effluent from fossil-fuelled stations, some further treatment may be necessary.

11.3 Water requirements

The need to supply large quantities of water to thermal power stations only arises where assisted cooling systems are required at inland locations. These water requirements are considerable:

- approx. 2.7×10^{-3} m^3/kWh for modern fossil fuelled stations
- approx. $4.2\text{-}4.8 \times 10^{-3}$ m^3/kWh for nuclear power stations

The higher water requirements for nuclear stations reflects their lower thermal efficiency (32-38%) compared with that of modern fossil fuel power stations (42%), and the fact that about 5% of heat input is dissipated via the flue stack in fossil fuelled stations. The favoured PWR nuclear reactor has a 32% thermal efficiency.

11.4 Land requirements

These can vary significantly according to the size, type and method of cooling of the thermal power stations, as well as to specific site conditions. New power station increments of base-load thermal generating capacity are likely to be added in anything between 600 MWe to 1350 MWe, although total capacity of new power stations in most cases is likely to be about 2000 MWe for fossil fuelled stations and between 1300 MWe and 2000 MWe for nuclear stations. (In France, there are several groupings of up to four nuclear power stations in one area, which will eventually give rise to a total installed capacity of around 4000 MWe.)

It is assumed that the very large majority of new thermal generating capacity will be either coal fired or nuclear capacity. In the table below, estimates of land requirements are given, assuming the average total generating capacity to be 2000 MWe.

Table 11.4.1

LAND REQUIREMENTS* OF 2000 MWe THERMAL STATION

Hectares	Coal Fired	Nuclear
Direct cooling	80-120	50-70
Assisted cooling	120-160	75-100

* Includes proportionate additional land required for switching stations, etc.

It is expected that a very large proportion of new coal fired power stations will require assisted cooling. For nuclear stations, the availability of suitable coastal sites will vary depending on the country concerned. Thus, for some Member States, e.g. France and the U.K., a large proportion of nuclear capacity added over the next 20 years will be of the direct cooling type; for other EEC countries, e.g. Germany, Belgium and Holland, it is likely that most additional capacity will require recirculating assisted cooling systems. Coastal sites for direct cooling will be increasingly hard to find in the 1990's, and so the proportion of direct cooled thermal stations will gradually decrease.

In considering the need for new generating capacity, it should also be recognised that about 25-30% of currently operating generating capacity will be retired before 2000. Many of the existing sites will not be suitable for the new larger thermal power stations, so that new sites will have to be found over and above the net additional estimated capacity requirement. However, the overall amount of land required for this purpose is unlikely to exceed the amount of land freed up by the closing down of old stations, although it may be that some of this replacement capacity has to be built in more environmentally sensitive areas.

12. IMPACT OF THE ALTERNATIVE ENERGY STRATEGIES

The previous sections of Part B have described the various impacts which can arise through the production and consumption of different primary and secondary energy sources. It also developed, where possible, a number of 'pollution or resource use coefficients', which related the emission of the different pollutants and the requirement for certain resources to the production/consumption of the particular energy type. By applying these 'coefficients' to the alternative energy strategies, it is possible to compare the environmental impact of the different strategies out to the year 2000.

12.1 Nuclear power

12.1.1 Accident risk

The chances of a major accident as discussed in Section 8.3 are directly related to the total reactor-years of operation under the various strategies. Based on an average capacity load factor of 62% (as was assumed in Section A of the report), and an average reactor size of 900 MWe*, Table 12.1.1 below gives the reactor-years of operation for the alternative energy strategies by the year 2000.

Table 12.1.1

NUCLEAR REACTOR-YEARS OF OPERATION

Energy Supply Strategy	Installed Capacity, GWe			Reactor-years operation
	1976	1985	2000	-2000
1A Maximum nuclear	18	102	292	3800
2A Maximum fossil fuel	18	90	175	2800
3A Max. renewables/ high nuclear	18	102	253	3500
4A Max. renewables/ high fossil fuels	18	90	131	2500
1B Max. conservation/ high nuclear	18	102	230	3000
2B Max. conservation/ high fossil fuel	18	90	129	2450
3B Minimum fossil fuel	18	102	225	3000
4B Minimum nuclear	18	90	90	2200

* This figure will be somewhat high for the low nuclear strategies and so will slightly underestimate the reactor-years operation for these strategies.

Even if no improvement were to be achieved in safety standards between the low and high growth nuclear strategies, then the chance of a core melt-down having a fatal result for the public is only 1.7 times more likely to occur by the year 2000 in the maximum nuclear growth strategy than in the strategy of minimum nuclear expansion (this minimum case assumes no further nuclear capacity is added beyond that already now operating, under construction, on order or firmly planned). By the year 2020, by which time the majority of stations built in the 1975-1995 period would be at the end of their lives, this ratio between the high and the low strategies would be 2.4 times.

Taking the chances of an accidental radioactive release causing death as $1\text{-}10^6$ to $1\text{-}10^8$ per reactor year (see Section 8.3), then it can be seen that the risk of such a death occurring by the year 2000 in the maximum nuclear strategy is around 1 in 350-1 in 35,000, compared to 1 in 600-1 in 60,000 in the minimum nuclear strategy*. These figures are similar to the risk of death to a member of the public being caused by single year exposure of smoking ten medium tar cigarettes a day (1 in 4000) and by a traffic accident in the UK (1 in 800) (46).

Three points emergy from this analysis:

i. the fatality risk to the general public associated with the operation of nuclear power stations is very small; these risks compare favourably with those associated with other primary and secondary energy source production (47);

ii. over the next 20 year period, the risk of such an accident from the Community's minimum nuclear programme becomes comparable to the fatality risks in a single year from events which have come to be accepted by the general public as facts of modern day living; the reaction to a fatal nuclear accident, however, is certain to be of quite a different order of magnitude to that from a traffic accident, and could have repercussions on the short to medium term development of nuclear power;

iii. The risk of an accident does not differ very greatly between the high and the low nuclear programmes by the year 2000.

..2 Nuclear fuel cycle operation

1. Enrichment and fabrication

The number of uranium fuel enrichment and fuel fabrication plants required to support the nuclear programmes is, of course, dependent upon whether or not Member States enter into contracts to purchase fuel

---

* Note that safety analysis in this area so far makes no distinction between the risk of an accident from, say, 20 reactors operating for five years, and five reactors operating for 20 years.

in enriched form of as 'yellowcake'. It is also difficult to be specific about what capacity these plants might eventually be. Those currently operating, under construction or planned for the Community, are of the order of 400-600 tonnes/year of enriched uranium (2.5-3% U_{235}). However, it seems likely, particularly with the URENCO centrifuge process, that final capacities might be of the order of 1000 tonnes/year output, and possibly even more.

It is probable that an increasing proportion of the total uranium fuel required for the high nuclear strategies will be enriched and fabricated within the Community. On this basis the number of these plants expected to be in operation by the year 2000 is given in Table 12.1.2 overleaf, assuming an 85% plant load factor.

2. Fuel reprocessing

As discussed in Section 8.2.4, the requirement for irradiated fuel reprocessing plants is also dependent upon the policy adopted by the nuclear industries towards their radioactive wastes, and on whether or not they eventually embark upon a programme of building Fast Breeder Reactors. Nevertheless, it would seem likely that eventually a proportion of the waste will be processed, particularly in the high nuclear expansion strategies.

Again, it is difficult to be precise about the eventual capacity of reprocessing plant. The COGEMA oxide fuel reprocessing plant at Cap de la Hague is currently 400 tonnes/year capacity, and will be doubled by 1981. A second plant is also being built which will have an eventual capacity of 1600 tonnes/year. BNFL in the UK is awaiting the outcome of a planning enquiry decision * to build a 1200 tonne oxide fuel reprocessing plant. In Germany, DWK-KEWA have a 1400 tonne/year project at Karlsruhe for the late 1980's. Table 12.1.2 overleaf projects the number of reprocessing plants that would be required to reprocess all the irradiated fuel generated under the alternative energy strategies by the year 2000, assuming 2000 tonnes/year design capacity, and an 80% load factor. It is possible that in the high nuclear options, individual plant reprocessing capacity may be even higher than 2000 tonnes/year, and that the number of plants would be confined to 4-5. The upper limit to plant capacity will, however, be partly determined by the ability to discharge low-level liquid waste.

---

* Permission has since been granted.

Table 12.1.2				
MAXIMUM NUCLEAR FUEL CYCLE PLANT REQUIREMENTS				
Strategy	Installed nuclear capacity/assumed capacity	Enrichment 100t/year	Fabrication 1000t/year	Reproces-sing 2000t/year
1A	292 GWE	10	10	6
2A	175 GWe	6	6	4
3A	253 GWe	9	9	5
4A	131 GWe	5	5	3
1B	230 GWe	8	8	5
2B	129 GWe	4	4	2
3B	225 GWe	8	8	5
4B	90 GWe	3	3	2

From this comparison it can be seen that, while the low nuclear strategies (4A and 4B) would require only a limited amount of reprocessing plant capacity, and the option would theoretically exist to forgo building this capacity until 1990, at high levels of nuclear power capacity (1A, 3A, 1B, 3B) there would be strong pressure to add additional nuclear fuel reprocessing capacity before a satisfactory testing period for the initial plant(s) had elasped. From a security point of view, it would seem appropriate to minimise the number of sites for reprocessing nuclear fuel. Therefore, in the high nuclear strategies, there may be a case for even larger capacity reprocessing plants.

12.1.3 Wastes

1. Radioactive gaseous effluent

If there were no restriction on the discharge of active gases, then the total discharge of Kr-85 over the period to 2000 would be 0.7 to 1.2×10^9 Curies. Without reprocessing, this emission would be 600 times less. Given the half-life of this isotope is just under 10 years, the accumulated level of radioactivity from this source would be rather less. However, it is clear that, if reprocessing of fuel is undertaken, it would be necessary to collect and store the krypton-85 for a suitable period, before release into the atmosphere. This would raise the cost of the reprocessing operation.

The corresponding total load for tritium (half-life just over 12 years) by the year 2000 would be 4.6 to 8.0 x 10^6 Ci, compared to natural levels of tritium of 100 x 10^6 Ci.

2. Solid and liquid wastes

The total amount of solid and liquid wastes that would be generated under the alternative energy strategies is set out in Table 12.1.3. overleaf. These figures are based on LWR reactor fuel reprocessing.

The following points can be made in respect of Table 12.1.3.

(a) the accumulated cladding and high level wastes by the year 2000 are about 40 times the volume stored by the end of 1975 in the maximum nuclear strategy, and about 25 times in the case of the minimum nuclear strategy;

(b) while the volume of vitrefied high level radioactive waste is not very great (equivalent to a cube 13 metres in height) its energy content is still considerable and would require a substantial amount of specialised cooled storage capacity, before being consigned to final safe disposal;

(c) seen in these terms, the difference between the minimum and maximum nuclear strategies assumes some significance; even if no further capacity were added, the amount of high active waste generated in the maximum nuclear strategy would eventually be 3 times that of the minimum strategy;

(d) with the volume of radioactive fuel cladding material generated by the year 2000, sufficient specialised fuel storage facilities will have to be found for this intermediate radioactive waste.

(e) if full reprocessing of irradiated fuel takes place, and none of the recycled plutonium is used as reactor fuel (either in FBR's or thermal reactors),a maximum of around 1000 tonnes of plutonium will have been accumulated by the year 2000; in the maximum nuclear case, around 75 tonnes/year will be produced in 2000 compared with 22 tonnes/year in the minimum nuclear strategy.

Table 12.1.3

SOLID AND LIQUID WASTES (1) FROM NUCLEAR POWER GENERATION ARISING BY 2000

Strategy	Total depleted Uranium(2)	Recycled Uranium(3)	Recycled Plutonium(4)	Inactive Wastes		Low Level Solids		Low Level Liquids		Cladding Wastes		High Level Wastes	
				solid	liquid	Vol	Acty	Vol	Acty	Vol	Acty	Vol(6)	Acty
	10^3 tonnes	10^3 tonnes	10^3 tonnes	10^3 tonnes	10^3 tonnes	$10^3 m^3$	10^6Ci(4)	$10^6 m^3$	10^6Ci(5)	$10^6 m^3$	10^6Ci	$10^3 m^3$	10^6Ci
1A	932	114	950	186	233	48	1.0	1.0	10-20	12.0	684	23	110,000
2A	685	83	700	135	170	35	0.7	0.8	8-16	8.8	504	17	80,000
3A	855	105	875	171	214	44	0.9	0.9	9-18	11.0	630	21	100,000
4A	640	76	625	126	158	31	0.6	0.7	7-14	7.9	450	16	70,000
1B	794	96	750	159	198	38	0.8	0.7	7-14	9.4	540	20	85,000
2B	600	73	615	120	150	31	0.6	0.8	8-16	7.7	441	15	70,000
3B	794	96	750	159	198	38	0.8	0.7	7-14	9.4	540	20	85,000
4B	548	67	550	111	136	28	0.5	0.6	6-12	6.9	396	13	35,000

(1) Assumes all waste is reprocessed.

(2) From enrichment plants.

(3) Based on high burn-up LWR fuel.

(4) Takes into account decay of short-lived radionuclides.

(5) Based on current allowable discharge levels at Windscale reprocessing plant in U.K.

(6) In concentrated form.

12.1.4 Proliferation and terrorist threat

Broadly speaking, the proliferation and terrorist threat are increased in proportion to the amount of nuclear fuel waste being moved and re-processed. From Table 12.1.2 it would appear, therefore, that the potential risk by the year 2000 would be 2-3 times as great in the high nuclear expansion strategies 1A, 3A, 1B, 3B, than in the low level of nuclear expansion strategies. The risk of theft of re-processed fuel would be extremely small in the minimum nuclear strategies 4A and 4B, since the number of processing plants could be restricted to 2-3 in the entire Community, and their development could proceed at a measured pace allowing time for design and security procedure improvements to be built into the system as operating experience was gained. A faster expansion programme would allow less time for such a process to occur. One of the results of this could be that greater police powers and personal security checks would be substituted for technical innovation in security measures in plant operation and fuel transport.

Nuclear proliferation risk cannot be eliminated if uranium ore is being imported and nuclear fuel enriched in the Community. There is undoubted room for improving the supervision of these activities by the International Atomic Energy Agency and EURATOM. However, the possibility of nuclear proliferation arising from the existence of nuclear power plants themselves in Member States needs to be kept in perspective in relation to other possible means by which proliferation could occur, e.g. the development of a civil nuclear power programme in non-nuclear weapon countries is likely to be a considerably greater threat in this respect.

12.2 Atmospheric pollution

12.2.1 Thermal pollution

Table 12.2.1 below gives estimates for thermal emissions from direct combustion applications and thermal electricity generation for the alternative energy strategies.

Table 12.2.1

THERMAL POLLUTION UNDER ALTERNATIVE ENERGY STRATEGIES

	kWh x 10^{15}								
	1976 actual	Increase by 2000 1A	2A	3A	4A	1B	2B	3B	4B
Direct fossil applications	8.9	4.0	4.7	3.2	3.6	2.9	3.0	1.8	2.1
Power stations	1.9	3.3	2.4	2.8	2.3	2.3	2.1	2.3	1.9
Total	10.8	7.3	7.1	6.0	5.9	5.2	5.1	4.1	4.0

In the low economic growth scenario, thermal pollution by the year 2000 will have increased by an estimated 3.9 x 10^{15} kWh, i.e. about 50% of the level shown for strategy 1A. The high economic growth scenario would increase the thermal emissions by a further 12% over that shown in 1A.

The following points can be made:

i. even in the worst case, total thermal pollution from the Community will only have increased by about 65-70% which would imply that man-made heat emission was still only about 0.03% of the natural radiation balance;

ii. because of the greater waste emission inherent in the high nuclear strategies, there is greater thermal pollution in the maximum nuclear strategy than in the maximum fossil fuel strategy;

iii. the share of total thermal pollution from waste heat from power stations increases from 17% to 25-28%;

iv. waste from thermal power stations will increase by 2-2.7 times by the year 2000; only if this heat emission is concentrated is it likely to have a measureable impact on the local environment.

12.2.2 Gaseous pollutants and particulates

The estimated emissions of sulphur oxides, nitrogen oxides and particulates under the alternative energy strategies are shown in Table 12.2.2. This 'worst case' assumes that no additional control measures are introduced, combustion technology remains the same, and average sulphur levels of oil fuel remain constant; because of a larger element of imported lower sulphur coal, average sulphur levels in coal are assumed to be 25% lower by 2000.

Table 12.2.2

GASEOUS POLLUTANT AND PARTICULATE EMISSIONS UNDER ALTERNATIVE ENERGY STRATEGIES

In 10^6 tonnes/year		← Increment →									
	1976	1985 Strategies		2000							
	Actual	"A"	"B"	1A	2A	3A	4A	1B	2B	3B	4B
SO_x	16	4	2	4	8	3	8	3	6	2	7
NO_x	5	1	1	1	2	1	1	1	2	1	2
Particulates	3	-	-	1	2	1	2	1	1	-	1
CO_x (x 10^3)	3	4	4	4	5	4	5	4	4	4	4

The high economic growth scenario would have the effect of raising the emission level of gaseous pollutants to those shown for the high fossil fuel strategies (2A, 4A, 4B). If the alternative low GDP growth postulate were realised, then gaseous pollutant levels would show little increase over those indicated for 1985.

The following points can be made with respect to Table 12.2.2.

1. $\underline{SO_x}$

In terms of SO_x (of which SO_2 is by far the largest component), it can be seen that substantial differences result according to the energy strategies adopted. If no programmes for further abatement are implemented, SO_x emissions in the high fossil fuel strategies could rise by around 40-50% by the year 2000 over 1976 levels. In the more limited expansion of fossil fuel consumption, the increase would be some 20-25%. In both cases coal combustion would account for the majority of this increase (particularly after 1985), most of which would be emitted from power stations.

For the most part, the current selective SO_x emission standards and the use of tall stacks should prevent the episodic build-up of ground level SO_x concentrations in the Community. However, there will be areas where topographic and meteorological conditions inhibit dispersal, and for these areas the high fossil fuel strategies will necessitate more stringent SO_x emission control standards, if ground level pollution from this source is to be prevented from getting any worse.

It should also be noted that the high fossil fuel strategies would imply a considerable potential increase in the export of SO_x from Member States if no further abatement measures are introduced. The means to limit SO_x emissions at current levels from increased oil consumption can almost certainly be achieved from refining crude oils of lower sulphur content, e.g. from the North Sea. Since most of the increased SO_2 emissions from coal combustion comes after 1985, there is good reason to believe that, in the case of industrial use, fluidised bed combustion should be able to provide effective gaseous effluent control by that time. For many coal-fired power stations, installation of flue-gas desulphurisation equipment may be the only available means of control, although in some cases, low sulphur coals may be substituted. By the 1990's, fluidised bed techniques should also be developed for power station boilers. Finally, it is worth pointing out the noticeable effect of maximum conservation policies in constraining SO_x emissions; adopting a high nuclear growth strategy would reduce potential emissions in the year 2000 by about 20%.

2. NO_x

On the basis of no additional control, increased emissions of NO_x are also evident for each of the strategies, up to 40% in the high fossil fuel strategies. Most of this increase comes from stationary sources and, as noted in Section 9.7 can, in some areas, cause an equivalent quantity of 'acid rain' to be formed as sulphur oxides . However, as fossil fuel combustion control techniques improve and with the introduction of fluidised bed combustion, so it should be possible to prevent overall NO_x emissions from deteriorating significantly from their current levels.

Of more concern is the potential increase in ground-level NO_x emissions from mobile sources. In this respect, gasoline combustion is much the most important and this is estimated to rise by 40-50% by the year 2000. Again, improved engine combustion and better air/fuel ratio control techniques (recirculating exhaust-gas systems would appear particularly promising) offer the most likely means of control. Catalytic control methods require the availability of lead-free gasoline, and as yet have not achieved proven reliability over a period of time.

3. Particulates

Growth in emission of suspended particulate matter is of principal concern in urban areas, and is likely to occur after 1985 unless stack-gas arrestment is yet further stepped up. Without further control, this increase would amount to about 60% in the high fossil fuel strategies, whereas opting for a high nuclear growth can reduce the emission level in the year 2000 by around 20%. Some additional improvement (10-20%) in precipitate or performance should be possible without a large increase in costs. The introduction of SO_x emission control techniques after 1985 will also further reduce fine particulate emissions.

4. Carbon dioxide

The annual emission CO_2 from the Community is forecast to increase by 60-65% in the high fossil fuel strategies, and by 40-50% in the high nuclear strategies. This increase will be small in terms of the total world increase in man-made CO_2 emission, and it is probable that by the year 2000, the Community will be responsible for not much more than 15% of total world man-made CO_2 emission.

12.3 Water pollution

All the strategies envisage the maximum exploitation of indigenous crude oil and coal resources, and, therefore, no difference exists between them as regards water pollution. As mentioned, accidental release of crude oil to the marine environment is the main cause for concern.

12.4 Solid waste

Forecast solid waste generation according to the alternative energy strategies is given in Table 12.4.1 below.

Table 12.4.1

SOLID WASTE GENERATION UNDER ALTERNATIVE ENERGY STRATEGIES

Million tonnes/ year	1976	◄---------- Increment		------------------							
		1985		2000							
	Actual	"A"	"B"	1A	2A	3A	4A	1B	2B	3B	4B
Coal mining	170	(25)*		-----------			(30)				-----------
Coal combustion	90	10		30	75	40	70	25	55	25	60

* ( ) indicates reduction

It can be seen that, in spite of some increase in forecast coal output from the Community, the actual level of waste generation is expected to fall somewhat. This is a function of the type of new coal seams being mined in future years compared to those currently exploited. Overall, this reduction will offset the increase expected from additional coal combustion in all but the high fossil fuel cases. However, for those areas where new mines are being opened up, and where it is likely that a number of new fossil fuel power stations will be located, the increase in solid waste could prevent difficult disposal problems unless the area is well endowed with landfill sites. Solid waste generation will be exacerbated if wet limestone scrubbing techniques are introduced to control sulphur emissions as this will produce a sludge of about 3 times the weight of the fly ash on its own. Fluidised bed combustion, on the other hand, would only increase the weight of solid waste by about 20%.

12.5 Water requirements

Considerable additional water requirements will be generated in meeting future energy needs. As can be seen from Table 12.5.1 overleaf, most of this increase results from thermal electricity generation. In the low nuclear strategies (2 and 4) it has been assumed that 50-60% of additional nuclear capacity will be directly cooled by sea water, whereas in the high nuclear strategies (1 and 3), this proportion is postulated as 30-40%. In the case of new fossil fuelled generating capacity, 90% is assumed to require recirculating cooling systems.

In 1976, fresh water cooling requirements represented about 37% (excluding Italy which has a very high hydro-electric use which distorts the total) of total Community water abstraction. On the assumptions taken, the estimated increase in freshwater requirements are relatively small (8-14%). This essentially is a reflection of the fact that the 1976 thermalpower station abstractions contains a proportion of direct cooling use. Since direct cooling takes in 50 times as much water as recirculating systems, the projections can be seen to be highly sensitive to this factor. And although most utilities envisage very little increase in direct cooling, the rise expected in cooling water requirements, could still place a considerable strain on certain water courses, particularly in summer. In some areas, it may be necessary to use mechanically assisted draught recirculating cooling, which reduces the water needs at the expense of thermal efficiency of the station. The greater requirements in the high nuclear strategies are a function of the lower thermal efficiencies of PWR stations than fossil fuelled power stations.

12.6 Land requirements

By far the largest land area needs of future energy supply and conversion plants will be generated by thermal power stations. In assessing the number of new thermal power station sites likely to be required it was assumed that overall Community system load factor for thermal stations would be 50%.(This compares with 46% which is the maximum that has so far been achieved , and with increasing contribution from renewables electricity generation, it is unlikely that overall system load factor will significantly exceed this 1973 level.) Forecast land requirements under alternative energy strategies are shown in Table 12.6.1.

Table 12.5.1

WATER DEMAND UNDER ALTERNATIVE ENERGY STRATEGIES

$10^9 \times m^3$/year	1976 (1)	Increment									
		1985		2000							
		'A'	'B'	1A	2A	3A	4A	1B	2B	3B	4B
Thermal power stations	6.0-7.0 (1)	1.3-1.5	1.4-1.6	4.5-5.0	2.4-2.8	3.7-4.3	2.6-2.9	3.1-3.9	2.4-2.7	3.1-3.7	2.5-2.7
Coal gasification plants				0.2	0.5	0.2	0.5	0.2	0.3	0.2	0.3
Total	40 (approx)	1.3-1.5	1.4-1.6	4.7-5.2	2.9-3.3	3.9-4.5	3.1-3.4	3.3-4.1	2.7-3.0	3.3-3.9	2.8-3.0

(1) Approximate estimate based on 1973 Member State statistics.

Table 12.6.1

SITE AND LAND REQUIREMENTS (1) UNDER ALTERNATIVE ENERGY STRATEGIES

	1976	←——— Increment	———	———	———	———	———	———	———	———	———→
		1985		2000							
	Actual	'A'	'B'	1A	2A	3A	4A	1B	2B	3B	4B
Coal gasification plant - sites (2)	-	-	-	9	18	9	18	7	11	7	11
- 10^3 hectares (3)	-	-	-	2-4	3-8	2-4	3-8	1-3	2-5	1-3	2-5
Thermal power stations - sites (4)	NA	60	55	200	140	165	135	145	115	145	110
- 10^3 hectares (5)	15-20	5	5	16-17	12	12-13	13	11	12	11	12
Oil refineries	12-16	-	-	-	0.5	-	0.4	-	-	-	-
Total - 10^3 hectares	27-36	5	5	18-21	15.5-20.5	14-17	16.4-21.4	12-14	14-17	12-14	14-17

(1) Net requirements.
(2) Assumes 2.3×10^9 m^3/year plant with coal input of 5.5 million tonnes/year.
(3) 160-440 hectares/plant. Variation depends upon buffer area required (48).
(4) Assumes average power station size of 2000 MWe.
(5) 50-150 hectares/station depending on station type and cooling methods - see Section 11.4.1.

The following points can be made:-

i. while the number of large coal gasification plants forecast to be in operation by 2000 is not very great, their land requirement can be seen to be considerable; local site factors will determine the degree of land buffer required;

ii. the number of new power station sites required by 2000 varies between 110-200, i.e. some 4.5-8 new stations per annum; by the 1990's, there may be considerable difficulty in finding suitable * new sites, particularly in the high nuclear strategies;

iii. even under the minimum electricity growth projection, there is a need to find 110 additional sites for thermal power stations over the 1975-2000 period;

iv. except in case 1A, where nuclear power expansion is maximised, there is greater land requirement associated with the high fossil fuel strategies than with high nuclear strategies;

v. in some situations there may be a trade-off between land area and the amount of cooling water that can be made available, i.e. through the building of additional assisted cooling facilities.

12.7 Impact of renewable sources of energy

Maximising renewable sources of energy has little overall impact on the environment by the year 2000. Little appreciable reduction in the impact of nuclear or fossil fuel production and use is achieved.

The impact of the renewable sources of energy themselves can be obtained by comparing those energy strategies (1A, 2A, 1B, 2B) in which 'renewable' energy sources (including CHP and heat pumps) are not given special priority and contribute 65 million tonnes o.e. by the year 2000, with those strategies (3A, 4A, 3B, 4B) in which renewable sources are developed at the maximum rate considered possible to yield 155 million tonnes o.e. by the year 2000.

In the low 'renewables' expansion strategies, it is assumed that one large new tidal barrage would be in operation by 2000, there would be some 40-80 kilometres of wave machines, and about 1-1.5 GW of wind power capacity installed, equivalent to some 200-300 wind turbines. The remaining 4 million tonnes o.e. contribution would be made up of geothermal power, photovoltaic electricity generation and refuse derived fuel. The scale of solar and combined heat and power developments assumed for the commercial sector is insufficient to make a significant impact.

---

* Access to cooling water is a major criterion in influencing site selection.

In the high 'renewables' expansion strategies, wave power capacity is assumed to be 1.5 times that of the low renewables case; some 500 x 5 MW windmills will need to be installed in France, Denmark, Ireland, Germany and the UK; three major tidal power schemes will need to be built. In maximising solar energy use in the domestic and commercial sectors, a marked degree of housing and commercial building planning will be required to enable sun-orientation and housing density criteria to be met.

12.8 Conservation

The beneficial effects of reducing environmental impact by a policy of concerted energy conservation measures can be observed in previous sections by comparing the impacts of the "A" energy strategies with the equivalent "B" energy strategies. The most significant of these can be summarised as:-

i. a 15-25% reduction in nuclear wastes;

ii. a 10-15% reduction in potential SO_x emissions;

iii. a 5-15% reduction in potential NO_x and particulate emissions;

iv. a 10-15% reduction in overall thermal pollution;

v. a 10-15% reduction in CO_2 emissions;

vi. a 15-25% reduction in solid waste generation;

vii. a 2.5-10% reduction in cooling/process water requirements;

viii. a 5-10% reduction in total land use requirements.

These figures refer to annual figures for the year 2000.

12.9 Comments

From the assessment of the environmental impacts deriving from the alternative energy strategies postulated, the following general observations can be made:

1. A degree of additional environmental damage will result from whatever energy strategy is adopted.

2. A policy of maximum conservation of energy provides the most effective overall way of reducing the various impacts as well as being the economic means of achieving this end, taking into consideration that such investment is also lowering the Community's energy costs and reducing its imported energy needs.

3. While the exploitation of renewable sources of energy would not appear, on the whole, to result in serious impact, their contribution by 2000 does little to reduce the overall environmental consequences of utilising conventional sources of energy.

4. We do not believe that the health and safety risks associated with the operation of thermal nuclear power stations, even in the high growth strategies, are any worse than those of embarking on the fossil fuel alternatives. In some respects, e.g. atmospheric pollution, solid waste, land use, the environmental impacts from nuclear power station operation are less than those associated with fossil fuel strategies.

5. However, there are environmental arguments against a high rate of nuclear expansion from the aspects of the nuclear fuel cycle operation and the storage and final disposal of high level radioactive wastes. In the former, there are definite environmental and national security hazards associated with the movement, storage and reprocessing of irradiated fuel and very little operating experience has yet been achieved in these activities; in the latter, no satisfactory solution has yet been found for the final safe disposal of these wastes. A fast rate of nuclear expansion could force commitment to certain courses of action in these respects, before their acceptability had been sufficiently proven. This would be less likely to result from a more measured approach.

6. A high rate of nuclear expansion is also necessarily associated with a rapid expansion of electricity consumption. The environmental impacts and resource requirements of thermal electricity generation and transmission are considerable, and,in some Member States, the need to find new power station sites in the 1990's could provoke increasing opposition from the general public.

7. Increasing fossil fuel usage inevitably will generate some additional environmental damage. In the case of oil and gas, the most serious potential pollutants can be contained for a relatively low cost. For coal, such containment will, to a considerable extent, rely upon the application of techniques which, at the present time, are at a pilot plant development stage or, whose full-scale implementation is not yet widespread. However, there is reasonable prospect that by the late 1980's, when the need for increased coal consumption is likely to arise, that such atmospheric pollution control techniques will be available without adding substantially to the cost of coal. In some areas, a potential problem could be the disposal of solid waste or sludge from additional coal mining and consumption.

8. The potential increase by 2000 of those forms of pollution which have been identified as leading to possible global climatic change is either small in relation to the increased emission elsewhere in the world (CO_2) or does not represent a large increase against the natural flux (as in the case of thermal pollution).

APPENDICES

APPENDIX 1

INDUSTRIAL SUPPLY OF DELIVERED ENERGY TO INDUSTRIAL SECTOR - ALTERNATIVE STRATEGIES FOR YEAR 2000						
Million tonnes o.e.	2A	3A	4A	2B	3B	4B
LPG	14	14	14	14	14	14
Distillate oils	56	57	113	48	44	41
Fuel oil	118	113	73	83	94	74
Natural gas	75	75	75	94	94	94
Coal/coke	142	124	113	127	112	127
SNG	25	16	25	20	10	20
Electricity	85	99	85	86	88	86
Solar	-	17	17	-	17	17
Total	515	515	515	473	473	473

INDUSTRIAL SUPPLY OF DELIVERED ENERGY TO INDUSTRIAL SECTOR - LOW AND HIGH GDP GROWTH CASES - 1A				
Million tonnes o.e.	1985		2000	
	Low	High	Low	High
LPG	6	7	12	12
Distillate oils	30	41	42	57
Fuel oil	67	100	77	121
Natural gas	93	93	75	75
Coal/coke	87	90	116	150
SNG	1	1	16	20
Electricity	51	60	68	112
Total	335	392	406	547

APPENDIX 2

FORECAST OF SUPPLY OF PRIMARY ENERGY FOR ELECTRICITY GENERATION - BASE CASE

Million tonnes o.e.	1975	1985			
		1A/3A	2A/4A	1B/3B	2B/4B
Oil	61	83	93	74	82
Coal	109	128	134	122	130
Hydro	25	35	35	35	35
Nuclear	20	140	124	140	124
Natural gas	34	29	29	29	29
Renewables	-	1	1	1	1
Total	249	416	416	401	401

Million tonnes o.e.	2000							
	1A	2A	3A	4A	1B	2B	3B	4B
Oil	66	92	53	145	50	125	58	155
Coal	138	165	145	179	120	115	120	178
Hydro	35	35	35	35	35	35	35	35
Nuclear	400	240	346	180	316	177	309	124
Natural gas	10	10	10	10	10	10	10	10
Renewables	23	23	47	47	23	23	47	47
Total	672	565	636	596	554	535	579	548

APPENDIX 2 (cont)

FORECAST OF PRIMARY ENERGY SUPPLY FOR ELECTRICITY GENERATION - LOW AND HIGH GDP GROWTH CASES - 1A

Million tonnes o.e.	1985		2000	
	Low	High	Low	High
Oil	72	83	88	74
Coal	128	128	143	152
Hydro	35	35	35	35
Nuclear	97	140	163	400
Natural gas	29	29	10	10
Renewables	1	1	23	23
Total	362	416	462	694

APPENDIX 3

REFINERY FUEL CONSUMPTION										
Million tonnes o.e.										
1976	1985		2000							
	A	B	1A	2A	3A	4A	1B	2B	3B	4B
32	40	40	45	47	41	45	40	42	34	40

REFINERY FUEL CONSUMPTION - LOW AND HIGH ECONOMIC GROWTH CASES - 1A				
Million tonnes o.e.	1985		2000	
	Low	High	Low	High
	35	45	40	45

APPENDIX 4

SECTORAL ENERGY DEMAND - LOW AND HIGH GDP GROWTH CASES - 1A

Transport Sector Demand

Million tonnes oil	1985 Low	1985 High	2000 Low	2000 High
Gasoline	105	107	109	115
Diesel	41	47	48	64
Jet kero	17	18	22	25

Domestic Sector Demand

Million tonnes o.e.	1985 Low	1985 High	2000 Low	2000 High
Solid fuel	18	18	25	30
Natural gas	105	105	130	130
Oil	115	141	90	80
Electricity	10	10	10	31
CHP	1	1	10	10
Solar	-	-	25	30
SNG	1	2	5	17
Total	250	277	295	328

REFERENCES

1. Second Report on the Achievement of Community Energy Policy Objectives for 1985. COM(77) 395 Final, Brussels, July 29, 1977.

2. Report on World Energy Demand 1985-2020. Prepared for Conservation Commission of World Energy Conference, September 1977.

3. Energy Global Prospects 1985-2000. Workshop on Alternative Energy Strategies (MIT), 1977.

4. An Analysis of the Energy Balances for the U.K. and Western Europe to the Year 2000. R. Belgrave. A paper given to the Royal Institute for International Affairs, March 1977.

5. Emission Control of Motor Vehicles in Europe. R. Jublot and E. Goulley. Paper given to World Energy Conference in Detroit, 1974, Ref.6.1.12.

6. Energy Conservation in U.K. Dwellings: Domestic Sector Survey. F. Romig and G. Leach. Published by IIED London, June 1977.

7. World Energy Outlook. OECD, 1977.

8. Vorblatt zur Entwurf eines Gesetzes zur Einsparung von Enegie in Gelauden. Bundesministerium für Wirtschaft.

9. Energy Conservation Ways and Means. J. Over and A. Sjoerdsma. Future Shape of Technology Foundation, The Hague, 1974.

10. European Coal 2000. Association of Coal Producers of European Community (CEPCO).

11. The Light Water Reactor System and its Final Cycle. D. Gaussot et al. A paper delivered at the International Conference on Nuclear Power and its Fuel Cycle. Salzburg, 2-13 May, 1977.

12. The Choice of Thermal Reactor Systems. National Nuclear Corporation Ltd, published by U.K. Department of Energy, 1977.

13. New Scientist, December 1977, pg 714.

14. The Potential of Natural Energy Resources. J.D. Denton et al. CEGB Research, 1975, No.2.

15. Wind Power. Paper given by N. Sorrenson. British Hydro-Mechanics Research Station, Cambridge, September 1976.

16. Waves at Ocean Weather Ship India. L. Draper and E. Square. Trans. R. Inst. Nav. Archit., 1967, 109, pg 85-93.

17. CEGB 1975. Evidence to the Select Committee on Science and Technology, Part IV. Session 1974-75, London HMSO.

18. The Exploitation of Tidal Power in the Severn Estuary. Fourth Report from the Select Committee on Science and Technology, Session 1976-77, HMSO, 1977.

19. Deployment of National Resources in the Provision of Energy in the United Kingdom, 1975-2025. Watt Committee on Energy, August 1977.

20. Nuclear Power Issues and Choices - Alternative Energy Sources. Nuclear Energy Policy Study Group/the MITRE Corporation. Ballinger Publishing Co., 1977, pg 141.

21. District Heating Combined with Electricity Generation in the U.K. Energy Paper No. 20. HMSO, 1977.

22. Cancer of Insolation and Ice Age Problem. M. Milankovitch. Transl. of Kgl. Serbische Akad. Spec. Publ. 132 (1941) by Israel Prog. Sci. (1969), U.S. Dep. Comm. Clearing House Fed. Sci. Tech. Inf.

23. As Ref. 20, pg 200-202.

24. Man-Made Climatic Changes. Helmut E. Landsberg. Science, Vol.170, No. 3964, December 18, 1970, pg 1266.

25. Reactor Safety Study, (Wash - 1400). Professor Norman C. Rasmussen et al. MIT, October 1975.

26. Royal Commission on Environmental Pollution - Sixth Report Nuclear Power and the Environment. Sir Brian Flowers et al. HMSO, September, 1976, pg 113-114.

27. Design, Construction and Operating Experience of Demonstration LMFBR's. C.V. Gregory and D.C.G. Smith. Paper to be given at Bologna, 10-14 April, 1978.

28. Current Boiling Experiments in the Doureay Fast Reactor. C.V. Gregory and D.C.G. Smith. Journal of the American Institute of Electrical Engineers, 1978.

29. Economics of Recycling - Data Evaluation Volume One. Prepared for the Commission by Environmental Resources Ltd. To be published by the Commission in 1978.

30. Pollution - Environmental News Bulletin. Vol.7, No.11, November 1977.

31. A.B. Meinel and M.P. Meinel. Science, 155, 189 (1967);
F.E. Volz. J. Geophys. Res., 75, 1641 (1970).

32. Man-Made Climatic Changes. Helmut E. Landsberg. Science, Vol.170, No.3964, December 18, 1970, pg 1267.

33. Mauna Loa and Global Trends in Air Quality. L. Machta. Bulletin Amer. Met. Soc., 53, 5, 402, 1972.

34. Physics of the Air. W.J. Humphreys. McGraw Hill, New York,ed. 3, 1940, pg 587-615.
R.A. Ebdon. Weather 22, 245 (1967).
J.M. Mitchell, Jnr. In presentations to Boston meeting,AAAS, December, 1969.
R.A. Bryson. Weather misc. 21, 56 (1968).

35. *Long Range Transport of Air Pollutants*. Published by OECD, 1977.

36. Environmental Science of Technology, pg 1148-1149, December 1977.

37. *Summary and Critical Analysis of Studies regarding the Cost of Desulphurisation on a German National Basis*. W. Brocke and N. Haug. Paper No. Ea.7. to Seminar, 1975.

38. *Vegetation Injury from the Interaction of Nitrogen Dioxide and SO_2*. D.R. Tingey, R.A. Reinert, J.A. Dunning and W.W. Heck. Phytopathology, 61, 1506-1511, 1971.

39. *Combustion-Generated Pollution*. J.M. Beer et al. Science Research Council, 1976.

Survey on NO_x Emissions from Stationary Sources and Techniques for Reducing these Emissions. P.C. Blokker. Stichtung CONCAWE, March 1975.

40. *Energy Alternatives - A comparative Analysis*. Science. Public Policy Programme. University of Oklahoma, May 1975.

41 *Automotive Emission Regulations and their Impact on Refinery Operations*. CONCAWE Report No.10, 1977.

42. *Pollution - Environmental News Bulletin*. Vol.4, No.9, 1974.

43. *Tidal Power and the Environment*. T. Shaw. New Scientist, 202-206, October 1975.

44. *Environment Outlines - Tidal Energy in France*. R. Gilbrat. Development Forum, September/October 1974.

45. *Nuclear Power - Issues and Choices*. Nuclear Energy Policy Study Group/ MITRE Corporation. pg 141-143, 1977.

46. *Royal Commission on Environmental Pollution - Sixth Report - Nuclear Power and the Environment*. Sir B. Flowers et al, pg 77. HMSO, September 1976.

47. *Nuclear Power - Issues and Choices*. Nuclear Energy Policy Study Group/ MITRE Corporation, pg 240-241.